NATURAL RESOURCE HANDBOOK
by **John Tomikel, PhD**

ISBN 978-1482609936

In the United States we are used to thinking in terms of "individual freedom and rights." Asian cultures think in terms of the rights of the community or society taking precedent over those of an individual. In some instances, we put the welfare of society above that of the individual when we condemn property for such uses as schools, highways, hospitals, or for national defense.

As our population increases there will be more demand for food, homes, roads, and schools. We will be forced to make decisions concerning which lands to preserve and which will be used to supply those necessities.

Contents
1. Natural Resources
2. Air Resources
3. Water Resources
4. Soil Resources
5. Food Resources
6. Human Resources
7. Forest Resources
8. Rangeland Resources
9. Wildlife Resources
10. Wetland Resources
11. Marine Resources
12. Mineral Resources
13. Energy Resources
14. Harmful Substances
15. Waste Disposal
16. Future Prospects
17. The Mother Earth Society

1. NATURAL RESOURCES

A natural resource is anything obtained from the earth that is used by humans to satisfy needs or wants. Conservation is the wise and cautious use of the resource.

The human population is increasing so rapidly that it can safely be said we are running out of resources. Countries, such as the United States, use more than their share of resources. The United States has about five percent of the world's population and consumes about one third of the world resources each year. We use eleven times the world average energy, six times the steel and four times the grain. A small population increase in the United States has repercussions around the world.

If we tried to raise the world standard of living to the level of the United States it would be an impossible task. In order to equalize the world resources we would have to decrease our consumption by more than seven hundred percent. We, of course, are not likely to do that.

Resources can be divided into renewable and nonrenewable. Forests and rangelands are renewable resources. Minerals are not renewable, once gone, they will be gone forever. Although resources such as soil are renewable, it takes such a long time to build a soil that they are considered almost nonrenewable.

We can conserve the nonrenewable resources by recycling, reclamation and substitution but eventually we will run out of them. Even the renewable resources might be in danger since we are not practicing sustainable management. We should be cutting our forests at the rate of forest growth but we are cutting the forests at a greater rate than that of growth. As a result, many places on earth where forests once flourished are now denuded. The environmental consequences of this have been dramatic.

America has been named the "throwaway society". We use resources at a tremendous rate and put them into landfills or pollute the earth with them. We should adopt a philosophy of a sustainable society where our use of resources is limited to the availability and conservation of those resources. This would require sacrifice on our part and a definite reduction in our throwaway mentality.

The conservation movement in the United States started about a hundred years ago (1908) when President Theodore Roosevelt and his Secretary of the Interior Gifford Pinchot invited civic and scientific leaders to form a National Conservation Commission. This resulted in the first Natural Resource Inventory and set up conservation agencies in all the states. Prior to this meeting, Congress had passed the Federal Forest Reserves Act (1891) which was an attempt to put federal forests off-limits to development and the Lacey Act (1900) which made it illegal to transport wildlife across state borders.

During the years of the Great Depression when a large number of Americans were unemployed, President Franklin D. Roosevelt began to rebuild America and give employment to individuals through such programs as the Civilian Conservation Corps which constructed bridges, built roads and started flood and fire control programs in our forests. The Soil Conservation Service was established in 1935. It dealt with the soil erosion and conservation involved with dust storms of the Great Plains. Other projects during this time period included the creation of the Works Progress Administration which worked on roads and city infrastructure. One of the biggest accomplishments was the North American Wildlife and Resource Conference of 1936. It started a tradition which lasts to the present day.

World War II (1941-1945) took the country's mind off conservation and many practices were put on hold as the country exploited its soils and minerals in the war effort. The country was in a state of catching up after the war. This war mind-set lasted into the 1950s.

The present conservation movement began with the publication of Rachel Carson's book *Silent Spring* in 1962. It detailed how the use of pesticides was poisoning the earth. The public became involved when the book brought out the fact that we were not only poisoning the earth but ourselves as well.

After the publication of *Silent Spring* the influence of industry and its military colleagues over Congress was, if not broken, at least weakened. Environmental concerns swept the nation. New legislation included the Wilderness Act 1964, Clean Air Act 1965, Solid Waste Disposal Act 1966, Species Conservation Act 1966, Wild and Scenic River Act 1968 and National Environmental Policy Act 1969. We became a nation of environmentally concerned citizens.

The following list of environmental legislation identifies the areas in the environment of concern. The environmental concern of the country brought forth the Noise Control Act 1972, Ocean Dumping Act 1972, Federal Insecticide, Fungicide and Rodenticide Control Act 1972, Federal Water Pollution Control Act 1972, Marine Protection, Research and Sanctuaries Act 1972, National Coastal Zone Management Act 1972, Endangered Species Act 1973, Clean Water Act 1973, Forest Reserves Management Act 1974, Safe Drinking Water Act 1974, Toxic Substances Control Act 1976, Federal Land Policy Management Act 1976, Second Clean Air Act 1977, Surface Mining Control and Reclamation Act 1977 and the Endangered American Wilderness Act 1978.

Ninety-five percent of all the scientists that have ever lived are alive today. This has great promise for the future since science is dedicated to understanding our earth and improving our lives. Scientists today support the ecological approach to resource management. Ecology is the study of the relationships between organisms in the environment. The philosophy behind the ecological approach states that **humans cannot damage one part of the environment without harming other parts of it.** Each species has its ecological niche, that is, a place where it fits into the ecosystem (contraction for ecological system). The population of all species living and interacting in an area at a point in time is a community.

The ecological approach to resources takes into consideration the impact of an activity on the total plant and animal community in that environmental area. With this in mind, when we build a dam we take into consideration not only its economic uses but also its uses in water and soil conservation, its recreational uses and its impact on life forms that will be destroyed and enhanced by the construction of the dam. We now look at a forest, not only as a source of timber, but as a source of beauty, flood control, erosion control, recreational area and the habitat of wildlife.

If an environment is undisturbed over a long period of time an ecological system will develop where all the plants and animals exist in a state of balance. They might get out of balance for a time but eventually the system will return to a state of equilibrium. When humans have interfered with the equilibrium we have had to deal with the unbalanced system and found that it was not in our best interests to leave a system out of balance if we can correct it.

In nature, everything seems to flow in cycles. The hydro logic cycle rains on earth which flows into the ocean which evaporates to form clouds and eventually returns to earth as rain or snow. In the nitrogen cycle, plants take nitrogen from the air, improves the soil with it and eventually the nitrogen goes back to the air. There are also carbon and phosphorous cycles. There is even a rock cycle. Since these are cycles it cannot be stated accurately where the beginning point is on the cycle wheel.

Closely related to the ecological approach in the study of the environment is the concept of biome. The biome is a large terrestrial community that is easily recognized by its peculiar plant and animal associations. Thus we have biomes of tundra, coniferous forest, deciduous forest, rainforest, grassland, savanna and desert. Each of these has its own conditions of soil development and association with a particular climate,

Communities of plants and animals are always in a state of change. They move through a series of successions. For instance a grassy area such as a golf course in northeastern United States will eventually, if left alone, grow into small shrubs and then into trees and a climax forest. As the changes take place, the plants and animals will shift, and populations will change according to the stage of succession. Various successions can be recognized in all phases of human progression.

We are a part of nature, not apart from nature. Where humans have challenged nature, humans have always lost out.

A Quick Classification of Natural Resources
Essential Resources - 1. human beings 2. air 3. water 4. soil
Renewable Resources - 1. animals 2. plants 3. bacteria 4. marine
Nonrenewable Resources - 1. metals 2. fossil fuels 3. fertilizers 4. stone

2. AIR RESOURCES

Air is a mixture of gases. We hardly think about air since it is such a crucial part of our existence. If we are deprived of air for more than five minutes we suffer adverse effects such as brain damage.

Lean air, that is clean dry air, consists of two major gases, nitrogen (78%) and oxygen (21 %). The three major gases of the remaining one percent are water vapor, carbon dioxide and argon. There are also minor amounts, almost traces, of helium, methane, hydrogen, ozone, and a variable amount of human created chemicals.

There are also tiny particles of various kinds that are suspended in air. Most of these are so small they cannot be seen even with an average microscope. These are aerosols and there may be as many as five thousand of these suspended particles per cubic centimeter of air. Atmospheric aerosols include pollen, dust, salt from ocean evaporation and smoke from both natural and human created sources.

The troposphere extends to an altitude of about eleven miles at the equator but diminishes to about eight miles in altitude as one approaches the poles. Almost all of our weather takes place in the troposphere - clouds, winds, precipitation.

Temperature decreases with altitude in the troposphere. The decrease continues to the top of this layer until it begins to level off and slightly reverse itself. This boundary is the tropopause. Above it is the stratosphere.

The stratosphere is often divided into several layers depending on the conditions found there. Mesosphere is the name given to the layer where the temperature begins falling again. Above this is the thermosphere where the temperatures begin rising again.

An important layer of the stratosphere is the ozonosphere, a thin band of ozone circling the earth. In the stratosphere the layer is about twenty miles thick but it must be remembered that the air at this altitude is so thin if this layer was brought down to sea level it would only be as thick as ten sheets of paper. However thin, the ozonosphere protects us from ultraviolet radiation (UV) given off by the sun. This is the radiation that causes sunburn and skin cancer (malignant melanoma). UV radiation also causes eye cancers and cataracts. It also has some serious effects on our immune systems. It also damages plant cells that perform photosynthesis, the basis of all life on earth. So the ozone layer is extremely important.

Ozone at ground level is harmful in many ways. It destroys plastic and rubber products. It causes disease by affecting the moisture in the nose, mouth and respiratory tracts.

The ozone layer in the stratosphere is being destroyed by various chemicals created by humans, notably CFCs, chlorofluorocarbons. These are found in refrigerant gases and in certain types of foam plastics.

Above the troposphere we find the ionosphere which is an extension of the thermosphere. When sunlight hits this layer it sometimes creates ionization of the gases. This manifests itself in the formation of visible gases known as the aurora. Also these ionized gases form into layers which deflect radio and television waves back to earth. Until satellites, this bouncing of long waves was the main method of obtaining radio and TV reception.

Air is hottest at the equator where the sun's rays are most direct. It is coldest at the poles. This unequal heating along with the rotation of the earth creates wind patterns and storms over the earth. Generally in the tropics the wind moves toward the equator, in the middle latitudes it moves from the southwest and west and at the poles it moves away from the poles.

Air Pollution

Pollution is any change in air water or soils that affects human health or survival. This definition can be expanded to take in animals and plants but it stands to reason if humans are affected so are other animals as well as plants.

The most prevalent air pollutant is carbon monoxide created mostly by combustion of fuels. Some is added to the atmosphere by volcanic eruptions and earthquakes. Combustion of fuels also produces a wide variety of other pollutants, mostly carbon dioxide which is believed by many scientists to aid in heating of the earth in a process called the Greenhouse Effect. This Effect is believed to be occurring at a greater speed since humans have added tremendous amounts of combustion particles to the atmosphere. However, no firm evidence has yet proved the Greenhouse Effect to be a reality.

Sulfur oxides and nitrogen oxides are caused by combustion of fossil fuels, notably by vehicle fuels and burning coal. These are the primary gases responsible for acid rain. Sulfur oxides merge with water vapor to produce sulfurous acid and nitrogen oxides merge with water vapor to produce nitric acid.

Particulates are suspended particles of water or solids that can be seen by the unaided eye. These are mostly smoke and water vapor that is in the process of condensing.

Photochemical pollutants are those that begin as familiar gases such as those produced by automobile emissions but are changed once sunlight acts upon them. Photochemical pollutants include ozone, peroxides, certain hydrocarbons and aldehydes.

Radon is a naturally occurring radioactive gas which is constantly being expelled from the earth. If a house is built along the radon escape routes then it is possible the house basement can be filled with radioactive radon and pose a serious health hazard. Radon testing kits are readily available to monitor this gas if a homeowner should desire to do so.

Lead and heavy metals are put into the atmosphere by chemical processes and by burning such things as slick magazines. Heavy metals are all carcinogenic, that is, they can cause cancer.

Heat from industrial areas of cities is also considered an outdoor air pollutant. Noise is another form of air pollution and many cities have constructed walls to reflect automobile engine noises away from residential areas.

The major sources of outdoor air pollution are transportation 49%, stationary combustion sources such as electric power plant incinerators and home heating 28%, industry 13%, miscellaneous sources 7% and solid waste disposal 3 %.

The major gaseous air pollutants are carbon monoxide 50%, volatile organic compounds 16%, sulfur oxides 16%, and nitrogen oxides 14%. Volatile organic compounds include methane, formeldahyde, chlorofluorocarbons, halons and benzene. There are also secondary by-products in the atmosphere, especially photochemical smog and ozone.

Indoor air pollutants are considered more dangerous to humans than outdoor air pollutants since we spend most of our lives indoors. These include cleaning and polishing materials, formeldahyde leaking from plywood and other construction woods, oven cleaners and cooking odors, cigarette smoke, asbestos fibers, paint fumes, faulty heating systems and deodorants. Some of these can be removed by indoor plants such as spider plants and various species of ivy.

Effects of Air Pollution

Most air pollutants will cause headaches, asthma attacks, coughing, shortness of breath, and nausea. Prolonged air pollution will cause death.

Carbon monoxide reduces the ability of blood to carry oxygen and this slows down body processes, especially thought processes. CO poisoning starts off as drowsiness, headaches, nausea and eventually a coma and death. It is the interference with thought processes which keeps the victim from realizing what is happening. Some scientists believe that most auto accidents in congested cities are caused by impaired thinking due to excess carbon monoxide buildup during rush hour traffic.

Suspended particles such as those created by cleaning fluids can cause asthma and bronchitis attacks. Long term breathing of these can damage lungs and cause respiratory diseases, cancer and even death. Suspended sulfur oxides increase heart attacks.

Heavy metals and materials in gaseous form such as Polychlorinated biphenals (PCBs), insecticides, dioxins, and other volatile organic compounds lead to reproductive problems and cancers. Many of these directly affect genetic materials creating birth defects and mutations in animals. Male sperm counts in most areas of the world have been decreasing steadily in the last hundred years. Some scientists believe this is due to association with environmental pollutants.

The metal lead causes retardation in children. Lead is put into the atmosphere by burning lead containing fuels, paints and magazines with glossy prints. However, the reduction of lead in the atmosphere by restricting it in gasoline has caused a dramatic drop in this pollutant.

The impact of air pollution is most dramatic in forest areas. In the United States we find the forests in the northeast located at the tops of mountains and hills are dying back. The Canadian maple syrup industry has been reduced by 60 percent since 1950. More than seventy percent of the forests of Poland, Slovakia, Germany and England have been destroyed by air pollution.

The Clean Air Act

For years there have been air pollution disasters in the world. More than 20,000 people were killed in one pollution crises in London. In 1948 a temperature inversion over the town of Donora on the Monongahela River in Pennsylvania prevented pollutants from steel and chemical industries from escaping into the atmosphere. On the first day of the pollution alert, twenty people died. Before the end of the week, ten percent of the town's population was seeking medical attention. Researchers are still investigating this event and find that all along the Monongahela River Valley deaths and hospital admissions reached abnormal proportions.

For over a hundred years environmentalists have tried to get the United States Congress to enact pollution controlling legislation. They resisted, since congressman are very sensitive to political contributions from industry who argued that this would "cost jobs." In reality, their argument was based on profits.

A great social upheaval occurred in the United States during the late 1960s and early 1970s. The unpopular Viet Nam War was in progress and social and environmental activists were enraged. Rachel Carson published her book *Silent Spring* in 1962 in which she documented the unregulated use of pesticides was not only killing off birds but humans as well. Even today pesticides are the least regulated of all polluting chemicals.

It was short of miraculous that Congress passed the First Clean Air Act in 1972, more than twenty years after the Donora incident and ten years after *Silent Spring*. Until then, Canada was complaining to international organizations of the tremendous amount of pollutants going into their lower Great Lakes region from the industrial valleys of Pennsylvania and Ohio.

The Clean Air Act of 1970 stated that the Environmental Protection Agency will (1) identity pollutants (2) determine how these affect health and the environment (3) identity the sources of pollution, and (4) develop a suitable method for control of the pollution. Levels of pollution are determined by the input amount, space occupied by the pollution and the out-take mechanisms, such as gravity, which remove pollutants.

Human reaction to pollution is based on the threshold level which is a level of pollution below which no harmful effects are observed, or in other words a level above which harm occurs. The threshold level depends on time of exposure, concentration of the pollutant and the nature of the pollutant. Radioactive thresholds vary with the type of radiation.

The exposure of a person to pollution is measured in a dose which is the concentration of pollution times the time of exposure. People working with radiation wear a badge which measures the daily dose they receive.

Outdoor chemical air pollution is usually a mixture of pollutants. These are often synergistic which means that one pollutant will increase the harmful effects of another, such as cigaret smoking coupled with asbestos and drugs with alcohol.

Another type of dangerous air pollution occurs when there is an increase in vehicle exhaust emissions. These gases change their chemical nature when exposed to sunlight. These photochemical changes are always aggravating to eyes, nose, throat and lungs.

Congress finally addressed the problem of auto emissions, acid rain producing gases and the prohibition on the manufacture of certain chemicals such as chlorofluorocarbons, halons, methyl chloroform and carbon tetrachloride. The pesticide DDT was banned ten years after the publication of *Silent Spring*. Despite clean air regulations urban ozone levels have actually increased since 1970. Most all states now require vehicles registered in cities to have an emissions inspection.

There are many more provisions needed in the clean air policies of the United States. We need to have stricter laws concerning municipal trash incinerators. Presently many of these put dioxins, lead and mercury into the atmosphere.

Stricter laws are needed for vehicle exhaust discharges as well as airline pollutants. However, consumers have revolted when faced with the prospect of paying higher prices for less polluting fuels such as gasohol.

Current legislation allows a "credit" system for industries. Nonpolluting industries receive so many credits for their non-pollution. These credits can be transferred to polluting industries in order for them to avoid paying fines and other taxes on their pollution. This system is controversial indeed. Its intent was to keep pollution at a constant level without any appreciable increase. This avoids legislation that would decrease this pollution.

Somehow, we have to improve air quality on a global scale. However, it is difficult to tell a newly emerging industrial country with limited financial resources that they have to become environmental friendly.

In discussing the remedies for cleaning pollution we have to be aware that dispersion of pollution is not the same as cleaning it up. A large smokestack merely sends the pollution on to the next town or the next country.

Some interesting tables.
STATES WITH HIGHEST PER CAPITA AIR POLLUTION.
1. Wyoming 2. North Dakota 3. West Virginia 4. Louisiana
5. Indiana 6. New Mexico
STATES WITH HIGHEST GREENHOUSE GAS EMISSIONS
1. Texas 2. California 3. Illinois 4. Ohio 5. Pennsylvania 6. New York
COUNTRIES LEADING IN WORLD GREENHOUSE GAS PRODUCTION
United States 18%, Russia 13%, China 9 %, Japan 6 % Germany 4 % Brazil 4 % India 3%
U. S. CITIES LEADING IN AIR POLLUTION AND THE AVERAGE NUMBER OF DAYS OF UNHEALTHY AIR. (five year average starting year 2000)
New York 330, Los Angeles 313, Phoenix 277, Seattle 229, Tucson 228, San Diego 272, St Louis 221, Sacramento 263, Las Vegas 214, Bakersfield 247, Nashville 197, Pittsburgh 247, Cleveland 192, Salt Lake City 245, New Haven 202, Fresno 241, Washington 199, Denver 239, Houston 188, Cincinnati 236, Raleigh 187

3. WATER RESOURCES

Fresh water is our life resource. We can only live about a week without it. Our bodies are almost 70 percent water and this needs to be replenished daily in order for us to stay healthy.

Fresh water is needed in many of our daily activities as well as in agriculture and manufacturing. Twenty percent of American food is produced on irrigated lands. It takes over one hundred thousand gallons of fresh water to produce one automobile.

Water is found in the ground, on the surface and in the air. It circulates over the earth in the hydrologic cycle. Water evaporates from the ocean and other water bodies to become water vapor in the air. Water vapor is lighter than any other air molecules and so it rises. As it rises it cools and eventually it rises to a cooling height that causes it to condense as small droplets. We see these in the sky or near the ground as clouds or fog.

Clouds move with the wind and as they rise further in altitude they become concentrated to the point where precipitation occurs. Precipitation may be in the form of rain, snow, sleet or hail. When precipitation falls to earth, it can do three things. It can form streams and run off (fluviation), it can evaporate back to the atmosphere (evaporation) or it can sink into the ground (infiltration). Once it is on the earth it is ready to evaporate and start a new hydrologic cycle.

Water vapor is also added to the air by plants in the process of transpiration. As water evaporates it takes heat from the surrounding atmosphere and this is why plants have a cooling effect.

The hydrologic cycle involves evaporation, condensation, precipitation, fluviation, infiltration and transpiration. As the process goes from liquid to gas, heat is taken on (heat of vaporization) and as the process is reversed heat is given up (heat of fusion).

Most of the water used in the United States comes from surface sources such as lakes, reservoirs and rivers. These depend mostly on surface runoff from drainage basins (watersheds). Because of pollution most surface waters need intense treatment to make them usable for domestic purposes.

Water infiltrating into the earth (groundwater) continues to move downward until it hits an impermeable layer through which it cannot continue its journey. As water builds up in this underground layer it fills in air spaces and voids of rock and soil until the layer is said to be saturated and this is termed the zone of saturation. The upper surface of this zone is called the water table. The saturated zone below the water table is called an aquifer. In order to pump water from this zone the well or pump must be placed below the water table. As water is withdrawn, a depression known as the cone of depression occurs in the water table around the well or pump. This causes more water to flow to that area and the well can continue to function.

The underground area above the water table is known as the zone of aeration since it contains air as well as suspended water particles. Most plants have their roots in this zone and many die when the water table rises and prevents them from ion exchanges between the root hairs and the underground gases.

When the water table meets the ground surface a spring or seep forms. The underground water can travels through porous rock or soil. This is an aquifer and it depends on permeability and porosity. Simply put, the hole spaces need to be connected in order for water (or oil or chemical pollutants) to move along.

If this occurs and the water table is above a drilled well then the water will flow under its own pressure. Such a well is said to be an artesian well.

About one third of the world has adequate fresh water supplies for the next fifty years while another third does not have adequate fresh water to meet today's needs. The other third suffers periods of drought on a regular basis and as their population increases they will be among those with inadequate supplies. The wars at the end of this century will be fought over water. Egypt has already declared its intention to go to war if Sudan builds projects on the upper Nile River. This situation can be found in over 150 river drainage basins around the world.

In the United States, the Great Lakes is the biggest reservoir of fresh water. Many demands will be made on this water. Several proposed projects to drain water to other regions have been met with opposition from the Great Lake States and Canada. One project to increase the flow of the Mississippi River during times of drought would drain water from Lake Michigan by way of the Chicago River.

Eastern United States has adequate water for the next fifty years at the present rate of consumption. Exceptions to this are the highlands of North Carolina and the Adirondack Mountains of New York State. Shortages of water are already felt in the vast area west of the Mississippi River. Severe shortages exist in the area known as the Southwest encompassing Arizona, New Mexico, Nevada, Utah, western Colorado, Wyoming and Southern California with its big metropolitan areas of Los Angeles and San Diego.

It is safe to say that any metropolitan area of more than a million people has a water shortage problem. New York City must supply 160 billion gallons of clean fresh water to its citizens every day. This is a tremendous task. This water ends up as waste and sewage water and it must be handled and disposed of in a sanitary manner. That is another tremendous problem.

Water Pollution

Water is scarce on a third of the earth's surface. Where it is available it is threatened by pollution which makes it difficult to treat and consume. Every major river of the world carries some kind of pollutant to the ocean. Every society that uses water must purity it if it wants to avoid disaster.

The major pollutant in most waters is caused by erosion and agriculture. As farmers till the soil much of it washes into streams and is carried away. This sediment is moved along by the river and it must be removed in order for the water to be used. The Mississippi River carries enough sediment to fill 60 railroad boxcars every hour, 24 hours a day. Sediment is also a product of ranching where overgrazing has bared the soil and exposed it to weather elements.

When fertilizers and animal manure enter the waterways these cause excess development of nuisance plants such as algae which clouds the water and prevents the growth of beneficial bottom plants (benthic). As the algae dies off it uses dissolved oxygen in the water which causes aquatic animals to die. This process is called eutrophication.

Untreated sewage causes pathogens which are disease bearing bacteria, parasites and viruses, to enter water systems. Pathogens also enter watersheds and streams from livestock and other animal wastes. Around the world about five million people die each year from diseases contracted from drinking water. Diarrhea is the leading cause of infant deaths in the world and almost all of the incidents come from drinking water. Other diseases transmitted by contaminated water include cholera, dysentery, gastroenteritis, hepatitis and typhoid fever.

Chemicals find their way into drinking water from the thousands of industries located around the world. Even the most sophisticated water treatment facilities usually treat for bacteria and have no reasonable way of removing chemicals. One pint of oil will contaminate a million gallons of water making it unfit for drinking.

When we identify the source of pollution we call that a **point source.** Usually these involve an industrial operation, coal mine, oil drilling, sewage discharge or a collection ditch or pond. In some instances the pollution comes from a large area such as a farming district and its not possible to identify. a single source. This is referred to as **non-point source pollution**. Non-point sources also involve city subdivisions, construction sites and highways.

Agriculture is responsible for sixty percent of stream and lake pollution. This comes in the form of sediments, fertilizers and pesticides. Agriculture produces more water pollution worldwide than any other human activity.

Streams will flush out their load of pollution providing they are not hindered by low water or damming. Lakes on the other hand are very slow to recover from pollution. While the river is flushing out, the cities which rely on it for water must shut down their intake pipes. A large oil spill on the Monongahela River in 1988 was monitored as it flowed into the Ohio and then down the Mississippi. The entire trip took three weeks. In another incident the Cuyahoga River of northern Ohio was so polluted with grease and oil in 1969 that it caught fire. Several wooden structures along its shores burned.

Eutrophfication mentioned earlier, shuts down and absorbs the water's natural oxygen recovery systems. This affects the biological oxygen demand (BOD) which is the amount of oxygen needed by aerobic decomposers to break down organic materials. It is also a measure of how much oxygen the pollution removes from water. BOD is different for different materials and involves time, temperature and volume of water.

Besides agriculture, excess nutrients are supplied by sewage treatment, pet wastes and phosphate detergents. Many states have banned the use of phosphate detergents and federal laws are tightening on sewage treatment and discharge.

Excess vegetation growth can be controlled by harvesting with a cutting machine. A modem cutting tool can cut water weeds to a depth of five feet. The cut weeds float to the surface where they can be gathered and taken to a landfill or composted.

Excess vegetation can be controlled by chemicals but these can also kill off beneficial plants. Some systems of oxygen enrichment have been used in several areas. This involves pumping air into the water to avoid oxygen depletion. This is an expensive way to control vegetation growth.

While streams can flush out their pollutants this is almost impossible for groundwater. Almost a third of our domestic water comes from groundwater sources. Usually small municipalities test for coliform bacteria, mainly Escherichia coli (E. coli) but they do little testing for chemicals.

The federal requirements for drinking water states that there should be no coliform bacteria in drinking water and less than 200 colonies of E.coli in a pint of swimming water.

About half of United States groundwater used for domestic drinking water is contaminated with chemicals that are a health threat. Contamination of groundwater from industry is especially high in New Jersey and Long Island. Contamination from agriculture is high in Florida and California where fertilizer and pesticide use is commonplace. The Environmental Protection Agency has discovered pesticides in groundwater in 39 states.

The Clean Water Act

The first clean water act (Federal Water Pollution Control Act) was passed in 1972. The series of Clean Water Acts requires the EPA to set up a nationwide system of water quality. Although it doesn't stop pollution it requires each polluter to provide an environmental impact statement before pollution can progress.

Cities and smaller municipalities are required to have at least secondary sewage treatment facilities and over 500 billion dollars has been spent meeting those requirements. However, most of the large cities have failed to meet final clean water standards. Cities such as New York City, San Diego and most cities along the coast of New Jersey still dump poorly treated sewage into the ocean.

The Safe Drinking Water Act of 1974 establishes maximum contaminant levels for any given pollutant. However, only 85 chemicals have come under regulation standards. There are over six hundred chemicals still being considered for regulation. Most of these are **carcinogenic** (cause cancer), toxic or cause disagreeable taste and odor. The Great Lakes is so polluted with some of these chemicals that states put warning notices on their fishing licenses. No one should eat catfish, lake trout, suckers or carp from any of the Great Lakes.

Privately owned wells such as those found in rural areas and trailer parks are not regulated by federal drinking water laws. However, even the regulated water suppliers of our nation have not been adequately monitored. Only six percent of the water law violators were prosecuted in 2008. Some cities were assessed fines of over half a million dollars. Unfortunately, they don't pay these fines and the EPA does nothing about it.

The Sewage Problem

As long as there are humans we will have to deal with human wastes. In the early days of civilization there was no treatment of human waste or sewage. Archeologists have stated that many ancient urban areas were destroyed by their own sewage build-up and the pathogens and diseases associated with it.

In days of old, in cities such as London, sewage was brought and dumped into the street for rains to wash away. Small streams were the transporters of raw sewage. In rural areas, much of it was buried or left in the open. Rural out-houses didn't appear in Europe until the late 19th Century.

Even today, large metropolitan areas such as New York City and San Diego dump lightly treated sewage into the ocean. Cities on the Mediterranean Sea such as Athens, Greece, dump more than half of their raw sewage directly into that body of water. Lagos, Nigeria, a city with over eight million inhabitants has no sewer system. More than half the people in the world do not have access to a sewage disposal system.

In most of the Lesser Developed Countries and the Newly Industrialized Nations where there are sewage systems they are of passive design. Sewage is collected into settling ponds and lagoons. Here the air, sunlight and microbes break down the sewage. After a month of settling, many of these municipalities treat the effluent with chlorine or some other chemical before the pond is cleaned. The collected material is usually taken to rural areas and used for fertilizer.

Most municipalities in modern industrial countries have sewage treatment plants. They also require septic tank systems in rural areas.

In modern sewage treatment systems, the human wastes from toilets, is moved through pipes to a central collecting point. Here primary treatment, which is a mechanical process, screens out debris such as plastic, sticks and cloth. The other solids settle to form sludge. The sludge is either gathered and taken to dump sites or is moved on for more treatment.

Secondary sewage treatment involves biological processes. These employ **aerobic bacteria** that break down the sludge. Another process is aeration and trickling. Sewage liquids are sprayed into the air which kills most of the associated bacteria. The falling water then trickles through a thick bed of ash, gravel, sand and rocks.

More advanced treatment depends upon the contaminants in the sewage. Mostly these involve separating heavy metals from the sewage and heavy chlorination. Some cities such as Palo Alto, California have found a wealth of heavy metals in sewage sludge. In their first year of metal recovery, Palo Alto sold the gold and silver from sludge for over 14 million dollars.

In some treatment system, **anaerobic bacteria** are added to sludge to decompose it. In this system the bacteria produce methane as a byproduct. This biogas can be used to heat buildings and even run vehicles and other equipment.

In many areas dry sludge is incinerated, but this causes air pollution and is especially bad when toxic substances appear in the sludge. If metals and toxins are removed, then burning sludge can be used to produce heat or electricity.

Probably the best solution is to disinfect the sludge and return it to the soil as fertilizer. Untreated sludge can be used as fertilizer in forests or in reclaiming old mine dumps. Usually in these areas there is little danger of groundwater contamination.

Many municipalities in the United States have storm water running into sewage water. When rains and melting snow is high in volume these overflow with raw sewage. This is a major cause of beach pollution.

In 2011, *The Big Thirst* by Charles Fishman was published. It was based on his years of research and its title tells its theme. The author states, "For Americans at home, flushing the toilet is the main way we use water. We use more water flushing toilets than bathing or cooking or washing our dishes or our clothes. The typical American flushes the toilet five times a day at home, and uses 18.5 gallons of water, just for that. What that means is that every day, Americans flush

5.7 billion gallons of clean drinking water down the toilet. And that's just at home."

As the population increases the human waste problem will increase with it. There are ways to handle the wastes in order to preserve our health and the environment but these come at a high price. There are also ways to lower the cost of sewage collection and treatment. Communities will have to seek these out.

Case Study:
The Sewage System of a Medium Sized City
ERIE, PENNSYLVANIA - pop. Served 130,000

The Erie Wastewater Treatment Plant is a secondary treatment plant having the design capacity of 69 million gallons per day. It averages, under normal conditions, between 30 and 40 million gallons per day from the city sewer system which is called the Municipal Flow. The secondary treatment process consists of the biological treatment of wastewater by utilizing many different types of microorganisms in a controlled environment.
The treatment is as follows:
1. Grit Screening: The municipal waste first enters the plant and passes through Bar Screens. These screens remove large rags, pieces of wood and plastic that could damage pumps and tanks. The screened out material is deposited in containers and removed to a sanitary landfill. At this stage, Ferric Chloride is added to the sewage water to precipitate out most of the phosphates that enter into the system.

Flow passes through the Grit Chamber, a large channel in which the velocity of the moving material is reduced sufficiently to deposit heavy inorganic solids but retains the organic material in suspension. The grit is removed by scrapers which travel along the bottom of the chamber. The collected grit is deposited in containers and removed to a sanitary landfill.
2. Primary Treatment: The organic flow of suspended materials is treated with an Anionic Polymer which forms a floc. This makes the larger and heavier solids (sludge) settle out in Primary Settling Tanks. These are long tanks with flight boards that travel across the top and bottom of the tank. When going across the top the boards remove floating solids such as grease. When returning across the bottom the boards collect the settled solids. The collected activated

sludge is then pumped to a wet well or holding tank for further processing.

3. Secondary Treatment: The combined primary effluents along with the activated sludge are then passed through Aeration Basins. Here the biological breakdown of organic material takes place. The primary effluent provides the food source and the sludge provides microorganisms for this process. The basins have a series of four passes containing a high level of dissolved oxygen which is necessary to sustain life of the aerobic organisms.

4. Final Clarification: The biomass generated in the aeration tanks is settled out in the Final Settling Tanks. The sludge that settle is collected and pumped back to the plant as activated sludge. The effluent from these tanks is then chlorinated for disinfection purposes and then the treated water is discharged into Lake Erie.

5. Sludge Removal: Not all of the activated sludge is returned to the aeration basins. Some of it is pumped to the Dissolved Air Flotation Units which are sludge thickeners. In these units, charged water, that is highly oxygenated water, and a cationic polymer are mixed with the activated sludge. The sludge forms a floc and the charged water causes it to float to the top of the tank. This thickened sludge is removed and pumped to the Wet Well where it is combined with the sludge from the primary settling tanks to receive further processing.

The sludge from the wet well is pumped to the belt press filters where cationic polymers are added and the sludge is passed through the presses. The resulting filter cake or sludge cake is conveyed to the incinerators for burning at temperatures of 1500 degrees Fahrenheit. The resulting ash is a clay-like substance which is deposited in a sanitary landfill.

Approximate Amount of Water (in gallons) Used in Various Activities

Car wash with hose 80, washing machine 60, ten minute shower 40, bath in tub 35, dishes by hand with water running 30, shaving, water running 20, automatic dishwasher 10, toilet flush 5, brushing teeth with water running 2 - new toilet engineering now enables a flush with slightly more than two gallons of water. In Japan hotels there are often two handles on the flush system, one labeled big flush and the other small flush.

Steps to Take in Saving Water: 1. Take short showers.   2. Don't run water while brushing teeth, shaving or washing up. 3. Repair all faucet drips. 4. Don't use the toilet as a wastebasket. 5. Wash only full loads in the automatic washer. 6. Don't use garbage disposals in the sink. 7. Clean sidewalks with a broom instead of a hose. 8. Water lawns and gardens in the early morning or late in the day.

4. SOIL RESOURCES

Humans depend upon soil for survival. Those countries with good soil and water resources prosper and those without do not. Some countries with limited food resources can compensate for this by producing products or selling resources which can be exchanged for food.

Soil is a mixture of decaying organic matter, mineral particles, living organisms, liquids and gases. It is a complex substance, an identifiable ecosystem.

Just as soils are complex, the creation of soils is a complex process. The development of soils begins with the physical and chemical disintegration and decomposition of parent materials. These materials are acted on by natural slightly acidic rain and atmospheric gases. Heating by the sun and freezing in winter creates physical disintegration. These processes are referred to as weathering.

Chemical weathering changes the composition of the original parent material and physical weathering makes smaller particles out of big ones. As the particles get smaller, more surface is exposed to chemical weathering. Once biotic life in the form of plants and insects take hold, the weathering and soil formation process is accelerated.

Soils have a texture and structure. Texture is the mixture of particle sizes and in soil classification these are sand, silt and clay. The best soil, loam, is a mixture of all three particle sizes. Structure is the way the particles merge. If clay is in layers in the soil then water cannot percolate through it. A layer of sand might let water infiltrate too deep and be unusable to surface plants.

There are many classification systems of soil. Most of these systems deal with the soil formation and their ultimate uses. An older system of soil classification based on formation process is still useful in understanding soils and soil formation. Under this system are three processes of formation.

Podzolization is the formation of soils under conditions of cool climates with adequate water. These soils form from disintegration of tree leaves and needles. These soils are acidic in nature and need "sweetening" (treating usually with pulverized and heated limestone) to make them productive. Podzol soils are found in eastern United States from the Great takes to the Gulf of Mexico.

Calcification is the process of soil formation created by dry conditions such as found in permanent grasslands. These soils are high in carbonates and need no sweetening. However they are deficient in moisture and irrigation is necessary to make them productive. These soils have names such as Chernozem (black earth), chestnut (red brown earth) and sierozem (desert). With adequate water these soils are very productive. Most of the world's wheat crop is raised on these soils.

Since moisture is scarce at the surface of the calcified soils the moisture comes to plant roots from below when it is not added above. Eventually the soils develop a crust of minerals on them (caliche) and if the crust continues to develop the soils are not productive. The crust can be flushed away but this takes more water in what is already a water scarce environment.

The third process, Laterization, takes place in tropical environments where there is a lot of heat and moisture. The only section of the United States with such soils is the Everglades at the tip of Florida. These soils, called laterites, are high in iron and aluminum. Due to high heat and precipitation laterites are easily destroyed once they are exposed to the weather.

Soils are studied and identified on the basis of a soil profile which is a cross-section of soil from the top of the ground down to the parent material, which is usually bed rock. The profile is broken into different horizons according to their distinct nature and stage of formation. The horizons originally were named A.B.C, and D but refinements in identification have added to this list.

The O (organic) Horizon is the top surface layer of material which consists of old and new vegetation (surface litter). Next in depth is the A Horizon (topsoil) formed in association with decaying organic matter, insects, gases, water and the activity of plant roots. This horizon is usually dark in color. Generally some of the nutrients in this horizon are leached or washed out. This physical removal of particulates is called eluviation.

Under the A Horizon is the E (eluviation) Horizon which is a layer of the A Horizon from which minerals have been removed by percolating water. This layer consists of sand and silt particles.

The B Horizon (subsoil) is next and consists of larger particles than the A Horizon and it is also an accumulation area for minerals washed out of the above layers. Farmers refer to the accumulation as "hardpan" since it is difficult to dig into. In some areas the hardpan keeps tree roots shallow since it is hard to penetrate and trees are easily blown over by high winds. The deposition of materials from the above horizon is illuviation.

Next down the profile is the C Horizon which is loose unconsolidated materials not reached by plant roots. It is also out of the zone of most surface weathering processes but is affected by physical weathering, by cracking from relief of stress such as when solid rock is brought closer to the surface.

The lower horizon is the R Horizon consisting of parent material!. The parent material may be bedrock, beach sand, glacial debris, volcanic lava or some other material freshly deposited by earth processes. This was once called the D Horizon but since "R" is accepted worldwide, soil scientists now use that designation.

Soil takes about five hundred years to develop into a natural productive entity. However, there are techniques to facilitate this development such as adding decayed materials to increase water retention and fertilizers to increase plant needed nutrients.

The U. S. Department of Agriculture has classified land according to its most effective use and capability. There are eight categories in this classification with category I and II being the most suitable for cultivation. Categories VII and VIII are not suitable.

The most usable land for agriculture is level land with good drainage and highly suitable for agriculture. There are few if any conditions that limit its use for agriculture. The classification changes in the best land when it begins to rise slightly and slopes increase and with slight drainage problems.

The land of the middle classifications is best used for pasture or orchards or forestry. The lands at the end of the classification system are best used for forestry and recreation. The land classified as VIII has steep slopes, almost no soil and lacks moisture retention capacity. It is best left alone.

Soil erosion is the biggest threat to the resource. This is the movement of soil from one place to another. Generally the movement is from the land to a stream and then to larger streams and ultimately the ocean. Once it is in the ocean it does not return.

As previously mentioned, it takes about 500 years to form an inch of topsoil in mid-latitude temperature zones. In undisturbed ecosystems soil is formed slightly faster than it is removed and there is an increase in soil depth. Once we engage in farming, grazing, building, logging and recreational activities the soil becomes loose and is removed by wind and water.

Water removes topsoil in sheets as it moves down a drainage slope (sheet erosion). This eventually runs into little furrows called rills, which takes the soil away faster. The rills turn into ravines and huge amounts of soil is carried away (gully erosion).

Streams are continuously eroding their banks and beds. The banks crumble into the stream and unless these are stabilized buildings and farmland are eventually destroyed.

Estimates of future world population state that the population of the world will double by the year 2050 and then level off at about eleven billion people. Today, two thirds of the world goes to bed hungry and half of these are in desperate need of continuous food supplies. Optimists believe that we can feed the forthcoming population by various heroic techniques. What these people don't seem to understand is that the soil resources of the world are continuously eroding away and by the year 2050, if present erosion trends continue the earth will only support about 3 billion people at most. This is less than half the present population.

Desertification takes place in lands on the fringes of deserts. Usually, these lands will support some people since they receive some precipitation and have thin soils. But as the populations in these lands increase more of the land is put to grazing. Once overgrazing occurs then the land is put to over tilling and when this occurs the land begins to erode with wind picking up dust and infrequent rain washing away the soil. The result, for all practical purposes, is a desert. This also can happen in dry forest areas as a result of over cutting vegetation for cooking fires and building shelters.

Desertification in Africa, south of the Sahara Desert is progressing at the rate of ten miles a year, that's a hundred miles in ten years. Severe desertification is also occurring in eastern Brazil, southern Peru, central Turkey, southwestern United States, southern Argentina, South Africa, southern Russia, western India and several areas of northern Australia.

Salinization is another soil problem This salt buildup occurs in areas near the ocean and where irrigation is constant. This is fresh water being used since ocean water will kill land plants, not grow them.

The United States Department of Agriculture's Natural Resource Conservation Service has many field officers teaching farmers how to save soil by using various techniques. This agency was formerly the Soil Conservation Service.

Contour farming was the biggest innovation made by the Service. By plowing fields in rows across the slope, rather than down, erosion can be reduced by as much as fifty percent. The water works its way slowly down the slope with each furrow taking in the soil removed from that above it. This is called "walking" the water off the land. This is easily understood when upland grain production is compared to that of the bottom land.

Strip cropping also helps stop erosion. This is planting crops such as com, grasses or beans in strips along the slopes or in the valleys. The grasses will capture the soil eroded from the open field crops.

Terracing has been widely used in the orient on steep hillsides. By building a wall on a hillside and filling behind it with materials and coating it with soil a flat area can be created for planting. The terraces will hold more water and let the water on it move slowly down to the next terrace.

Planting trees and shrubs as windbreaks (shelterbelts) is another soil saving technique. The windbreak not only keeps soil from blowing away but the reduced wind velocity will reduce the amount of evaporation taking place in a given field.

Many dry farming techniques have been invented. These involve drilling seeds into the ground without plowing, planting between rows of crops with other crops such as alfalfa between rows of corn, and heavy mulching. Mulching is putting vegetable matter over the soil to keep in moisture and retard erosion. A new plow, the chisel plow, will make a furrow about two inches wide. This is easily planted and covered with very little surface disturbance.

Unfortunately we need to use chemical fertilizers to keep up present crop production. These pollute waterways and cause diseases in many countries. Environmentalists are trying to get farmers to switch to organic fertilizers in the form of manure and green manure. Green manure is plowing a green crop under the soil where it will rot and increase water holding properties.

In many areas of the world, farmers put human sewage on their fields for fertilizer. This "night soil" is potentially dangerous in that much of it contains pathogens. People traveling in Third World countries are advised to wash and cook vegetables thoroughly before consuming them, if indeed, to eat them at all.

In a rural village in China we observed people simply going out into their fields and using it as a toilet. This, of course, fertilizes the soil and increases crop production. In this case the crop in the field was green beans. This practice is common in many parts of the world.

Another problem with food production is the heavy use of pesticides for weed and insect control. In countries such as Uzbekistan in Central Asia the use of chemical fertilizers and pesticides has caused tremendous rises in cancer and lung disease. Pesticides are the least regulated of all poison chemicals in every country.

But consider this: insects get one third of all crops planted in Third World countries and fifteen percent of crops planted in the United States. Other pests include fungus, rodents and wild grazing animals. In the United States weeds reduce crop yields by ten percent. Air pollution by ozone decreases American crop yields by another ten percent. Without pesticides these figures would be much higher.

Many farmers in the world are switching to a system of pest control known as Integrated Pest Management (IPM). In this system each crop and its pests are evaluated on the basis of ecology. A program is devised using a mix of cultivation and biological and chemical methods. Timing and sequence are important for maximum pest control.

Since it is almost impossible to eradicate all pests the theme of IPM is to reduce the pests to a level of low economic damage. Pesticides are applied only when the pests have become a serious threat to the crop.

IPM uses predator bugs such as ladybugs, strip cropping, crop rotation, green manure and genetically engineered crops as well as some other techniques. One of these other techniques is the use of a small vacuum which sucks the harmful insects into a bag where they are subsequently destroyed.

IPM requires expert training and education to be effective and United Nations training teams are busy traveling to all parts of the world to provide this training. If IPM was adopted worldwide there would be a major decrease in pollution from pesticides and fertilizers as well as a reduction in soil erosion.

Irrigation of Cropland

Irrigation of cropland accounts for sixty percent of the water used in the United States and about sixty-five percent worldwide. Almost two-thirds of that water is wasted. Most of this water is supplied by government agencies to farmers at discount prices. If farmers had to pay the true cost of this water then conditions might be different.

Most irrigation systems rely on gravity to carry water into furrows or ditches. More elaborate systems use pumps and various pipe devices to transport and release the water. Irrigation water is subject to evaporation, infiltration beyond crop root depths and runoff.

Various devices can be used to prevent infiltration, at least in the transport ditch, but little can be done to prevent evaporation and this is high considering that most irrigation is practiced in dry climates. Perforated pipes can be used to deliver water to each plant, thus cutting the loss of this precious fluid. This drip system delivers about eighty percent of the water to the crop and is the most effective method of irrigation. Spray systems can deliver water closer to the ground than sprinkler systems and in these about seventy percent of the water is utilized by the plants.

Worldwide about 120 acres are irrigated for every one thousand persons. This is about 16 percent of cropland and this produces one third of the world crop harvest.

Irrigated acreage peaked at the height of the Green Revolution (1978) and has been decreasing about four percent per year since then. Irrigation has not kept pace with population growth.

Every country in the world has some irrigated cropland. In the United States, California leads all other states in amount of water used for irrigation which is over thirty billion gallons a day. Idaho is second using twenty-one billion gallons a day. However, the irrigated water is sixty-two percent of California's fresh water use while it is ninety-two percent of Idaho's.

Most of the United States irrigated water is supplied at huge discounts by the federal government. Much of this is used to produce crops that are subsidized in other ways. As a result we are wasting water to produce surplus crops that must be sold at a discount. In the end, the American taxpayer loses at both ends.

Much of the existing irrigated land is losing productivity because of water logging and salt buildup. When drainage is not adequate the underlying water table rises causing water logging in many instances. In dry climates evaporation near the surface leads to an accumulation of salt which can ruin the land. \

The salt buildup or salinization affects about one-fourth of United State's irrigated land to some extent. The problem is severe in China, Pakistan, India and Mexico.

Over-pumping of groundwater for irrigation is another serious problem in many important crop regions. It is a problem in western United States, northern China, the Middle East and north Africa. It is critical in India's Punjab region which is the major supplier of grain to that undernourished country.

Much of the slowdown in irrigation is due to urban expansion where cities need more water at the expense of farming. As population increases, the situation will become more critical.

Environmental laws are having some effect in slowing irrigation expansion in the United States. California farmers lost some water supplies as a result of a 1992 law requiring much of federal water to be used to restore rivers, fisheries and wetlands. There is also a move to restore environmental integrity to San Francisco Bay.

An excellent example of the water dilemma is the situation in the area of the San Francisco Bay. Here the water resources are stretched to the limit among the needs of cities, agriculture and the need to protect the bay environment as well as the streams emptying into the bay. This region irrigates 200 different crops which make up 45 percent of the nation's fruits and vegetable production. It contains 120 different species of fish and supplies drinking water to 20 million people.

When water is diverted to irrigation and cities, it depletes the estuary of fresh water which increases salt content of the water and salinization of the underground water surrounding the Bay.

In 1995, the Environmental Protection Agency and representatives of the three water using groups drafted an agreement which satisfied all concerned. Each group did not get its demands but each group realized that some concessions were necessary for harmony to exist.

The arrangement established limits on how much fresh water would be diverted from the estuary to agriculture and cities and in which seasons. Cities would be affected most in dry years. The agreement was expected to return commercial and recreational fishing as viable enterprises. This has happened to some extent.

The big problem is the increase of population in the area. This agreement will need continuous monitoring and revision as will all water agreements throughout the world.

U.S. State Irrigation Use as A Percentage of All Water Use (2003)
Arizona 86, California 62, Colorado 91, Idaho 92, Kansas 83, Montana 96, Nebraska 71, Nevada 90, New Mexico 86, Oregon 87, South Dakota 68, Utah 83, Washington 70, Wyoming 91.

This table might be misleading or difficult to use since it does not entail the actual amount of water used. For instance the California use is 31 billion gallons of water, Idaho is 21 billion, Colorado is 12 billion. For the other states the amount is less than ten billion gallons of water. No matter how you look at it that is a lot of fresh water.

5. FOOD RESOURCES

Most scientists will agree that we have the ability to produce enough food to feed the world population. The problems of producing food and getting it to the consumer are many. About 90 million people are on the brink of starvation somewhere in the world while many countries have large surpluses of food and much food is going to waste. It is not simply a matter of distributing the food but many complex problems of politics and self-interest exist in food production.

Although humans eat any kind of meat we have found nine animal species to be the easiest and most desirable to raise for food. These are cattle sheep, goats, hogs, water buffalo, chickens, ducks, geese, and turkeys. In some sections of the world more specialized animal husbandry produces llamas, yaks, horses, reindeer, antelope, rabbits and a host of smaller animals, including dogs.

There may be as many as 70,000 plants with edible parts in the world that are being eaten today. However, many of them are obscure and less than a hundred are regularly eaten by people.

There are only four plants which really feed the world population. These are wheat, rice, corn and potatoes. These are produced in such great quantities that they take up more volume than all the other plant foods combined.

Potatoes are used near to where they are grown and do not enter into international trade to any significant degree. The three cereal crops on the other hand are significantly exported and imported between boundaries.

The biggest world producers of wheat in order of production are China, Russia, United States, India, France and Canada. However, not all of these countries are the biggest exporters and even though some of them are the biggest producers they also number among the biggest importers.

The wheat exporting countries in order of tonnage shipped are United States, Canada, France, Australia and Argentina. The wheat importing countries in order of tonnage received are Russia, China, Italy, Japan and Egypt.

The greatest rice producers of the world are China, India, Indonesia, Bangladesh and Thailand. The exporters of rice in order of tonnage shipped are Thailand, United States, Vietnam, Pakistan, China and Italy. The big importers of rice are Iran, Russia, Saudi Arabia, Brazil and Malaysia.

Not every statistic is without flaws. For instance, north China ships out wheat which is docked in south China. In reverse, south China ships rice which is docked in north China and these are counted as exports and imports.

Leading corn producers include United States, China and Brazil. The big exporters are United States, China, France, Argentina and Hungary. Big importers are Russia, Japan, South Korea, South Africa, Netherlands and Malaysia. Most corn is fed to animals such as hogs and cattle.

Grains, or cereals as they are called, have many advantages as a world food. They have a hard outer coating and are easy to store for long periods of time and they are easy to transport. They can be made into a variety of dishes. For instance, wheat makes bread, pastries, macaroni, spaghetti, pancakes. noodles and breakfast cereals. The chances are you had consumed wheat in at least three different forms yesterday.

The next food crops in order of production are barley, sweet potatoes, cassava, grapes, , sorghum, sugar cane, millet, banana and tomato. Cassava, or Manioc, and sweet potatoes are major root crops of tropical farms. We know cassava as tapioca.

Food is produced on small farms in most of the world but in the United States and rich countries the corporation farm has taken over production. These industrial farms involve large amounts of capital, that is, money. They usually grow only one crop from hybrid seed or specialize in raising one animal. Their equipment involves considerable operating and purchasing costs. And they usually have choice flat land on which to produce the food.

Corporation farms use chemical fertilizers and pesticides in large amounts. They are the major users of irrigation water. When they have a choice, most of them engage in meat production, especially chicken and beef. The small farmers of the world usually produce just enough food for themselves and a few neighbors. They have little money to invest. They own simple equipment and there is hand labor involved in production of meat and vegetables. They usually plant a variety of crops which feeds them throughout the year.

A major problem of small farmers who try to get into commercial production is their timing. If they produce wheat they have to sell it during the harvest to pay off loans. The price is low at harvest time. Then when winter comes and they need bread they buy their own wheat back at inflated prices. When it's time to plant they have to buy the seed at high prices. Thus, they are usually always in debt and cannot break the cycle.

Some small farmers are subsistence farmers. They just produce enough food for survival and when they have a bad year it means death or extreme hardship. Subsistence farmers also engage in hunting and gathering where it is feasible. Most of the farms in the rainforests of the world are subsistent. They are found in central Africa, along the Amazon River and on scattered islands in the Pacific Ocean.

A close relative of the subsistent farm is the intensive agriculture farm where hand labor is the key to production. These farmers are tied to the land. They do use fertilizers, mostly animal manure and they do get excellent yields on their small plots. Intensive agriculture is carried on in the northern Andes mountains of South America, China, India, Burma, Cambodia and Vietnam. There is also small agriculture intensity practiced in the back yards of Eastern Europe.

Plantation farming is a form of industrialized agriculture. The plantation grows cash crops usually from trees. These include bananas, coffee and cacao. Pineapples and sugar cane are also grown on plantations. Other non-food plantation crops are rubber, palm oil and cotton.

About thirty years ago scientists embarked on a crusade called the Green Revolution, by concentrating on developing new seed varieties, putting idle land under cultivation, making pesticides and fertilizers available to poor farmers and harnessing more water sources. This resulted in great strides in food production.

Because of the Green Revolution, China was able to feed its population and avoid famine, a historical event in that country. Pakistan and Thailand which were food importers became food exporters. Before the Green Revolution, half of the world population went to bed hungry every night. This number was reduced to one third, despite rapid population growth.

Mistakes during the Green Revolution were many. Fertilizer and pesticide pollution increased to dramatic proportions in some countries. Marginal lands were plowed and these increased soil destruction. Overgrazing occurred on the rangelands. Irrigation excesses caused distress to streams and underground water resources and increased salinization occurred in coastal and lake areas.

However, the fact remains, the world was able to feed a larger population. Techniques used in the first green revolution have been analyzed and world food organizations are now able to advise poor countries and their farmers on better methods of agriculture and preservation of the land.

The major users of pesticides are those engaged in agriculture 74 %, households 13 %, government 12 % and forestry 1 %.

About seventy percent of the grains produced in the United States is fed to livestock. Worldwide about 40 percent of all grain is fed to livestock, mostly chickens and hogs. It takes about four pounds of grain to make one pound of chicken, eight pounds of grain for one pound of hog and sixteen pounds of grain for one pound of beef.

We can feed more of the world population if we fed the grain directly to people and cut out the meat. It takes about 500 pounds of grain a year to keep a person healthy.

Today, seventy percent of all U.S. farmland is used in producing meat. Worldwide the number is thirty two percent. Feeding livestock involves huge amounts of fertilizers and pesticides as well as energy inputs. About forty percent of the world fish catch is used to feed livestock and fertilize crops.

The United States has a surplus of food each year. We get much of that surplus by giving government subsidies to corporation farms and to a lesser extent to individual family farms. This permits the corporation to sell the excess production to foreign buyers at a cost below production.

This "dumping" is met with hostility by farmers around the world. By subsidizing our farmers we compete unfairly with foreign farmers. As a result many countries have put high import taxes on American farm products. France is one of these. The U.S. retaliated by putting a high import tax on French wines. The French farmers marched on Paris, rioted and burned American flags. The French president went on national TV and reminded the farmers that France had taxes on American farm products and that many American soldiers from two world wars were buried on French soil.

The U.S. can produce rice cheaper than any country in the world, even cheaper than a less developed country. (LDC) To avoid disaster to its rice farmers Japan refuses to let American rice into the country. Rice farming is heavily subsidized in Japan where consumers pay higher prices for rice. They figure the higher prices are worth it in order to protect their basic farming industry. For a long time U.S. apples were banned from Japan but recently they have rescinded this ban. It is cheaper to send Washington State apples to Japan than to send them to New York City.

Rich countries have erected import taxes and other tariffs on food. Countries like the United States can sell food cheaper than poor farmers in other countries can produce it because of subsidies and mechanization. By selling food in poor countries, local farmers can't compete and they end up buying the cheap imported food themselves.

The same problem exists with humanitarian aid. By giving free food to starving nations we discourage their farmers from raising extra food since they can't sell it because organizations like the United Nations are giving it away. In the end the farmers line up along with the others for free food and the agriculture base of the nation is destroyed and the nation is constantly dependent on humanitarian aid.

The humanitarian aid is complicated by some governments wanting the food to be given to them and let them distribute it. This has not worked out well since these governments ripped off the food and sold it or gave it to their henchmen to distribute as they will. When humanitarian groups tried to circumvent the government in Somalia the distribution vehicles where hijacked,, presumably with government encouragement. In the month of July 2011, I noted at least one segment, every day, that the national news reported on the famine situation in African countries and the refugee camps. Bandits continually terrorized these camps.

Many countries subsidize their farmers in order to keep them in rural areas. Most cities of the world are overcrowded and by keeping the rural population earning a living it will slow down the staggering migration from rural to urban areas.

In the 1980s politicians in Nigeria ran for public office with the promise they would impose price ceilings on food. Since urban dwellers outnumber rural dwellers the urban candidates easily won elective office and price ceilings were imposed. Nigerian farmers could not make a profit at the set prices and simply stopped producing food for city markets. The government had to start importing food from neighboring countries. This caused a heavy drain on the national treasury and taxes were increased. In the end, price ceilings were dropped but the country has not yet recovered from this fiasco.

Simply said, poor countries and poor people do not have money to buy food and rich countries do not want to give the food away. They will sell food to humanitarian organizations who will distribute the food free of charge but these humanitarian organizations are operating on limited budgets and can't feed everyone in need. They also do not have the resources to combat political corruption which is rampant in most poor countries.

World Hunger

There are two types of hunger – hollow hunger where the stomach is empty and hidden hunger where there is food but it doesn't have the nutrition necessary for good health. A conservative estimate is that a third of the world suffers from hollow hunger and another third from hidden hunger. Two thirds of the world population goes to bed hungry every night.

When I was in Russia and discussed this topic with a geography professor he said that there was a third type which he called humdrum hunger. He said his diet in his childhood years consisted of nothing but potatoes. He was, of course, happy to have them. He said his mother had at least two hundred recipes for fixing potatoes.

About 45,000 children die **each day** from hunger related diseases. The main disease killing children is diarrhea which affects mostly children under five years of age.

Hunger and malnutrition kill about sixty million people a year. Undernourished people do not have energy to work and are likely to be affected by disease and this cuts down on productivity.

A diet deficient in vitamin A can result in blindness. A deficiency in B vitamins can cause nerve damage, deficiency in vitamin C can cause scurvy, deficiency in vitamin D can cause rickets. Iron deficiency is prevalent in many African women.

Two brutal diseases found in children of poor countries are marasmus, which is a wasting away of the body, and kwashiokor which makes the child lethargic and causes bloated stomachs. Marasmus causes the child to look old with wide eyes and shriven skin. Both are caused by diets low in calories and protein. Kwashiokor is also related to babies being weaned too early and deprived of mother's milk, a result of malnutrition in the mother.

Some areas of the world are nightmarish. In about thirty years India will have the largest population of all countries. India still continues to have a high birth rate despite its predicament. The average person of India is too poor to get enough food to meet basic nutritional needs. As their population soars their farm land suffers erosion, water shortages, and lack of fertilizer.

Although the cows of India, which are not killed because of religious beliefs, provide plenty of manure which could be used as fertilizer, the manure is often used for cooking fires or production of methane which is used to produce electricity in rural villages. Added to all of this is the fact that much of the irrigated farm land of India has been destroyed by salinization.

Several strategies for increasing food supplies have been advanced. More poor countries can get involved in fish farming and using trash fish by grinding them into fish patties. Non-traditional crops can be grown in lieu of the big four, crops like amaranth, dandelions, winged bean and fruits, such as mangos and breadfruits.

People can eat insects. There are many insects that are eaten regularly around the world. Insects are rich in protein but we have to get rid of our prejudices against consuming them.

Population limitation seems to be the key to a world free from want. Somehow the regions that produce an excess of food should get it to those regions that need food. To solve both of these problems will take extreme international negotiation and cooperation.

Food Resource Trends

Trends that shape our future are tracked by the Worldwatch Institute. In their annual publication "Vital Signs" most of the recent trends are negative. They note that grain yield per acre has dropped continuously at about three percent per year in the last ten years and is still dropping. Although this does not sound like much of a drop, it represents millions of tons. In the last ten years there was a one percent rise in production of grain each year. However, world population growth was one point seven percent (1.7%) which nullified the gains. Eighty percent of the world's people live on grain.

In China wheat production rose to record in 1978 when the communist leaders switched its farming to private enterprise. There was an increase in production equaling 85 percent in the next ten years. In the last ten years the increase in China's production has been 6 percent but overall the production had doubled since 1978. The population of China, though increasing, has been pretty much stabilized.

Decreasing trends are seen in production of all grain. Grain yields increased during the last twenty years because of increased fertilization and irrigation. Eventually the limit to which fertilizer could boost grain production was reached. Now world grain production is beginning to decline while the population continues to increase.

Irrigation is declining. Much irrigated land has become waterlogged. Salinization has affected 12 percent of all irrigated farm land. Over-pumping of ground water has depleted much of this resource. Also much of the water once set aside for irrigation has been diverted to cities where populations are increasing dramatically.

In California where much of the United States crops are produced, new laws cut back on irrigation water supplies. Most of the water must now be kept in rivers to restore fisheries and wetlands.

Fish farms produce about 16 million tons of fish and shellfish annually. Aquaculture produces ninety percent of all oysters marketed and one fourth of all shrimp. About two thirds of aquaculture involves inland water where carp, tilapia, trout and catfish are raised. The rest is in coastal marine waters where salmon, flounder, clams, oysters, crabs and shrimp are raised.

China produces half the aquaculture products of the world. India is second and Japan third. These countries produce 80 percent of farmed fish and seventy-five percent of all captured seafood.

Fish farming is a lot more involved than throwing fish into a pond and letting them reproduce. A fish farmer has to purchase supplies such as production equipment, antibiotics, hormones, vaccines, cleaning equipment and oxygenating materials.
The United Nations Food and Agriculture Organization (FOA) statistics indicate that the 17 major fishing areas of the world have reached their natural limits of reproduction. Nine of these areas are in serious decline.

Since the statistical base year of 1990, a few countries have increased their ocean catches, notably Russia and China. However, the coastal habitat of most countries have experienced overfishing, serious pollution and habitat destruction. More than fifty thousand Canadian fishers lost their jobs due to decreases in fish populations in their traditional waters. Water off the coast of Massachusetts is so polluted the United States banned fishing there. Fishing in polluted waters is a common practice and since the United States has very lax seafood inspection programs we are always in danger of buying contaminated seafood.

To insure fish in the future, ocean quotas will have to be set and enforced. The world per capita catch in 1988 was 43 pounds. In 1994 it was 39 pounds and in 2012 it was 34 pounds.

A summary of the fishing problem was given by the Earth Policy Institute:" The world catch of wild (marine?) fish per capita peaked in 1988 at 17 kg; by 2005 it was down to 14 kg. [Earth Policy] The fishing industry sends out 4 million vessels to catch wild fish, but stocks of the larger species are falling rapidly, so the industry works its way steadily down the food chain. Over the past 50 years, the number of large predatory fish in the oceans has dropped by 90%. Catches of many popular food fish such as cod, tuna, flounder, and hake have been cut in half despite a tripling in fishing effort."

The world faces declining seafood catches per person and rising seafood prices. There have been many conflicts over fishing grounds in the past. In 1995, Russian naval vessels forced several Japanese trawlers from fishing in the Sea of Okhotsk, a prime area for pollack, cod and herring. In that same year, the Canadian Coast Guard seized a Spanish fishing vessel found in its coastal waters and brought it into port. Incidents such as these will increase in the future.

There is some promise of increasing fish production through genetic engineering. So far, a super carp has been developed. It grows twice as fast as regular carp. And this new fish has been accepted readily by carp consuming people.

6. HUMAN RESOURCES

In 1930 there were 2 billion people on earth. By 1960 the number had risen to 3 billion and by 1975 there were 4 billion. The next billion was added by 1987 and in 1995 we had 5.7 billion. In June 2011 the world population was just short of seven billion. Population experts predict that the world population will increase to 11 billion around the year 2050 and then level off.

The world population is increasing by 350,000 a day, more than a million people are added every three days. If we are to give each of the new babies one glass of milk and a half loaf of bread each day, then it will take more than 25,000 new cows and 500 new acres planted in wheat each day. The world is decreasing in those commodities rather than increasing.

Most scientists predict that the increasing population will put a severe strain on earth resources. Wonder how they came to that conclusion? Poverty levels will rise and standards of living will decrease to miserable levels. However, other scientists believe we can feed this large population through technological advances and standards of living for most of the world will increase rather than decrease. They point to the fact that even though the population has quadrupled in the last century the world is no worse off than it was a hundred years previously. In fact, life spans have increased dramatically for average people in that period of time.

Whether you believe that there are too many people or not, you will have to agree that people are a resource. Other people provide the things you need to live and be comfortable. In most instances, other people provide you with food, clothing and shelter. They provide you with transportation, medical care and recreation. People are a natural resource in every sense of the word.

In 1798 Thomas Malthus published a book in which he stated that the plant world multiplied by arithmetic proportion 2, 4, 6, 8, 10 and the human world by geometric proportion 2, 4, 8, 16, 32. If this continued, humans would outstrip their food supply and there would be warfare, famine and pestilence in dramatic numbers.

Since the time of Malthus there have been famines, wars and pestilence which killed millions of people. However, the population continued to rise as advances in medicine and producing food have kept more people alive. Most scientists agree that the advance in population numbers is mostly due to a decrease in death rates.

Some countries want an increase in population, especially those where old people outnumber young people. For instance, in the United States, the Social Security System which pays out huge sums of money to retired citizens as well as disabled and dependent citizens, needs a steady flow of money in order to keep operating. Unless drastic changes are made in the payout there will not be enough young working people compared to retirees in 20 years in the country to keep the system working.

Most countries would like Zero Population Growth (ZPG) which is the number of births required for a population to continue replacing itself without increase. Worldwide, the rate is 2.1 but in rich countries the number is 2.2 since many women in these countries choose not to have children.

Many young couples say they would only like to have two children, that is, to replace themselves. However, a couple is two and if they have two children, there are now four where before there were two. It's a no win situation.

The number 2.1 is the fertility replacement rate. The fertility rate is an estimate of the number of children a woman between the ages of 15 and 44 will have in her lifetime. These rates are calculated on present births per thousand and female populations in a given country. Present rates for selected countries: Pakistan 3.2, Venezuela 2.4, Mexico 2.3, U.S. 2.06, U.K. 1.9, Australia 1.7, Canada 1.58, China 1.54, Germany 1.4, and Japan 1.2. Generally the fertility rates are the lowest in Europe and the highest in Africa.

The fertility rates have shown some changes in the last fifteen years. For instance, it was 5.6 in Mexico in 1995, 3.9 in India and 1.6 in Japan. The situation in Japan can be described as panic since their young women are not having children. Their immigration laws practically eliminate any newcomers to the country and they have to rely on their own people production to maintain social programs. They have one of the most homogeneous racial make-ups in the world.

Population studies are usually based on the crude birth rate (CBR) which is the number of births in a year based on a population of 1000. The crude death rate is the number of deaths in a year based on a population of 1000. The U.S. CBR is 15 and the CBD is 9. Since the U.S. Population is 312,000,000 we multiply this with the difference in CBR-CBD, six times 312,000 (not 312,000,000) and find the U.S. Population increasing by 1,872,,000 each year, at the present rate.

If we take the yearly increase per 1,000 (6) and divide it by 10 we get the percent of increase. If we divide that (.6) into 70 we get the years it would take to double our population (117). We should divide into 72 but 70 is easier to work with and gives us a good approximation. This number is based on the amount of growth at 1 % a year. If you put a dollar into a bank account at one percent a year it would take 72 years before you had two dollars in that account.

One consideration of the crude birth rate based on a population of 1000 is that in many instances half of that population would be male and half of the rest would be females who are not of reproductive age. However, the system gives us some idea of population dynamics and is accurate enough for strategic planning purpose.

However, the increase in birth rate is not the only factor affecting the United States population, there is immigration and emigration, the number of people coming into the country and the number of people leaving the country. One very comprehensive study of illegal Mexicans coming into the the U.S. was made in 1994. The conclusion was that Mexicans were illegally coming into the country at the rate of slightly more than one thousand a day. Studies indicate that this hasn't changed much in the last ten years. When this factor and other considerations is extrapolated the doubling of the U.S. Population is expected around the year 2055. Ironically, that is the year in which the Social Security fund is expected to disappear. The fund is supposedly solvent until 2040.

If the food supply remains static and the population doubles it does not take much mathematics to calculate that there is only half the amount of food available per person. That also means there is only half the amount of resources available per person.

If there are twenty of us sitting in a room and we each have a lunch of a cheeseburger and a milkshake and all of a sudden twenty people come into the room and we are forced to share our lunch, so we end up with half a cheeseburger and half a milkshake. Get it. How do we deal with it on a worldwide basis?

Throughout history increases in population have been managed in three ways which are (1) decreasing the number of people (2) increasing the food supply, and (3) limiting the number of births by birth control and family planning..

We have decreased the number of people within land areas through emigration, wars and genocide. In Rwanda, Africa in 1994, two different ethnic groups (Hutu and Tutsi) battled each other for four months for control of the land. When fighting ceased, over a million people had been slain. This reduced the Rwanda population by 12 percent.

The historic genocide of Nazi Germany and the Soviet Union need no elaboration here. The massive deaths of World War II had little impact on population. Despite all the deaths in wartime Japan their population was still larger after the war than before.

As you read this page, there are over fifty wars being fought somewhere in the world. The basic situation is a majority ethnic group trying to force a minority ethnic group to give up some of its land, its lives or resources. The resistance of the minority to do this leads to armed conflicts. When a minority group can break away from the majority it usually applies for admission as a country to the United Nations. If present trends continue the United Nations will double its membership in the next 50 years. Nations composing Chechens,

Kurds and many other groups cannot be held in minority positions any longer. In 2011 South Sudan won its independence from Sudan.

Attempts to increase the food supply have met with success until recently. In 1995, world food production decreased from 1990. There was a huge decrease in grain, vegetable, meat, fruit and seafood production. World relief organizations simply could not deliver enough food to the starving people of the world. Food production in 2010 was static compared to 2005 but the world population increased by half a billion.

Limiting population by birth control seems to be the preferred method of most countries. Sterilization is the most widely practiced form of birth control in the U.S. and in the world. For men the procedure is vasectomy and in women it is tubal ligation.

Contraception is the preferred method of birth control for many who may want to have children at a later date. This falls under two categories: physical and chemical contraception. With physical methods of contraception the sperm is prevented from reaching the egg by such items as inter-uterine devices, condoms, diaphragms and cervical caps. With chemical contraception, sperm is killed or a woman is prevented from ovulating. This method uses items such as the pill, spermicidal creams, skin implants, vaginal foams and steroids.

At the U.N. Cairo Conference on Population (1994) discussions of birth control took place. Even though this meeting took place many years ago it still is a useful study since it brought forth attitudes that are prevalent but unpublicized today.

Many plans were discussed. The Roman Catholic Church and many Islamic nations objected to much of the discussion focusing on family planning and rights of women. The Roman Catholic Church believes in only the natural rhythm method of family planning. That is, no artificial means of preventing the sperm from reaching the egg is acceptable. If the couple involved do not want to have children they calculate fertile periods by watching the menstruation dates of the woman. Half-way between these dates is the fertile period, usually 13 days after a woman begins menstruation. American Catholics, as a group, generally practice methods that are not church approved forms of birth control.

At the Cairo Conference statistics were distributed indicating the use of contraception by women in various countries. Some of these were (in percent) France 80, Sweden 78, South Korea 77, U.S. 74, China 71, Thailand 66, Japan 64, Peru 59, Mexico 53, Egypt 47, India 45, Zimbabwe 43, Chile 43, Kenya 33, Iran 23, Pakistan 12. Researchers concluded that more women would use contraceptives if they were available.

Scientists measure the population-land ratio by carrying capacity which is the maximum number of individuals of any species that can be supported over a long term by an ecosystem. Once the carrying capacity is reached, countries should try to attain zero population growth. This is a theoretical goal rather than a realistic one. Infinite growth in a finite system is impossible. Each day we get closer to the absolute limits of growth.

There are many reasons couples have large families. One very strong reason is cultural acceptability. People in many cultures have become used to large families. They continue to support this culture dictate even though the relation between poverty and population is evident to them.

More children are born to cultures where there is a lack of reliable contraception available. More than 90 percent of the women of Nigeria said they would use contraception if it were available.

When my married friends Joe and Peg visited her Irish roots Joe was shocked when he could not buy condoms in that country at that time..

In many cultures, men consider it unmanly to use condoms and the pressure on birth control is shifted to women. This is particularly true in Latin America and Africa.

The more education parents have, the less likely they are to have large families . Education also influences family income and those who have less children in a given area are better off financially.

Children in the work force is another factor in lesser developed countries. In countries such as Indonesia, Malaysia and Burma girls as young as twelve years old are employed in clothing factories. They are slaves in our modern world. Boys and girls as young as ten years old are employed in weaving enterprises in many central Asian countries.

As people migrate to urban areas they tend to have fewer children. Back on the farm children are an asset, in the city they are not productive and are mouths to feed. On the farm they can feed chickens, work on crops and bring in the cows. Older people can take care of many farm duties. In the city, old and young people are wards of those aged in-between.

The cost of educating and raising a child enters into family planning. Malthus believed that people should never have children unless they can prove that they can support them. It takes about five thousand dollars to raise a child for the first year in the United States. Throw-away diapers alone will cost $700 the first year.

Education in many areas such as Latin America is considered necessary and many parents make unbelievable sacrifices to send their children to elementary school. In Africa, more than eighty percent of women never attend school. At the Cairo Conference mentioned previously, the status of women in the world was given high priority. If women are employed and educated they will have fewer children.

Infant mortality rates have decreased worldwide but many cultures continue to have the same number of children per family unit as in previous ages. Seventy years ago it was common for one out of three children in lesser developed countries to die before the age of ten. People in these regions tried to have many children in hopes that some of them will make it to adulthood. As adults they would help to support their aging parents. Even though infant mortality has declined, some people seem to feel that it is still necessary to have large families. Some countries have reversed this trend. In 1965 the average woman in Thailand had 6.3 children. In 1987, the number had dropped to 2.2 Significant drops in fertility have occurred in China, Cuba, Indonesia and Tunisia.

Average marriage age has some influence on family size. This is assuming women do not have children before they marry. This is not the situation in the United States. In most countries marrying at a young age produces more children per family and marrying at a later time produces fewer children.

The average age for a women to marry in Ireland is 26 and a man is 32. In the United States it is 23 for women and 26 for men. In China, the government has recommended the Rule of 52. A man and woman contemplating marriage should have a combined age of 52 or more. China gives many incentives to couples that have only one child. They get first choice of government jobs, housing and food. This one child policy has led to the killing of many girl babies since Chinese men consider it unmanly not to have a son. The Chinese government started executing parents who kill their newborn daughters and this alleviated infanticide. Many people living in China West violate this one child policy without intervention since the government would like more buffer people in those areas.

When one considers the one child policy it is a shocking culture movement. It eliminates the terms brother, sister, cousin, uncle, and aunt.

Abortion is another method of limiting populations. Abortion occurs when a pregnancy is terminated before coming to term and the fetus is killed. There are an estimated 150,000 abortions a day worldwide. These are mostly in Lesser Developed Countries formerly called Third World Countries.

Family planning would eliminate the need for most abortions. In many countries such as most of those in Latin America, abortion is not legal. However, the abortion rate in many Latin American countries is higher than in places where abortion is legal. Roman Catholic Italy usually has the highest abortion rates in Europe.

The United States Supreme Court ruled that during the first three months of pregnancy abortion cannot legally be prevented by states. About one and a half million legal abortions are performed in the United States each year. About twenty percent of Americans believe abortions should be illegal. The confrontation between abortion proponents (Pro Choice) and abortion foes (Pro Life) has caused much hardship for both sides. The situation is heading toward a serious climax with Pro Life forces beginning to use terroristic tactics to shut down abortion clinics. There are about two thousand such clinics in the United States.

Migration of People

When conditions deteriorate for people they usually look about for more opportunity elsewhere. They migrate to other countries if they can be accepted. Many people migrate to these countries whether they are accepted or not. Recent history has seen thousands of people from Latin America and Asia crossing the United States border illegally. Many people consider legal immigration as well as illegal immigration as the biggest threat to the economic and social health of the United States.

Only a handful of countries will accept immigrants. Most countries are very protective of their borders. Some set up military units to either kill or frighten illegal immigrations.

There are about twenty million people fleeing their homelands every year in search of a place to live. Most of these are the result of wars or ethnic intimidation. More than half the population of Afghanistan was displaced in the recent wars there. Almost all of them have moved into Pakistan and Iran. These countries will tolerate them up to a point and then they will have to move elsewhere.

Refugees are people who flee their country because of fear of political, religious, or ethnic persecution or war. Today there are about four million refugees living in Europe, one million in Latin America and six million in the Middle East. The thousands of people living in camps for displaced peoples in Africa at this time can be added to the refugee problems of that continent.

The big movement of people, however, has been from rural to urban areas. Cities around the world are growing at an alarming rate as people abandon the land and look for a better life in cities, usually the capital cities of their country.

Migration to cities has led to the development known as the supercity or megacity. This is a city with a population of over ten million. Consider what it takes to get water, food, housing, clothing, medical care, trash pick-up, sewage disposal and electricity to such a place. It is mind boggling. Everything has to be brought in since the city is dependent on other places for its sustenance.

A good definition of the areal limits of supercity is based on its activities or function. Generally, one of the criteria in classification is the furthest distance a worker in the city is willing to commute.

Cities are population clusters of continuous built-up areas with the city boundary as the core. In this definition New York City would enter into New Jersey.

Present supercities would take in (order of largest populations) Tokyo, Mexico City, Sao Paulo, Seoul, New York, Bombay (old name), Osaka, Calcutta, Rio de Janeiro, Tehran, Buenos Aries, Cairo, Jakarta, Lagos, Manila, Delhi, Karachi, Los Angeles and Moscow.

People migrate to cities for many reasons including employment, housing, medical care and social programs. In their villages these may not be available.

If the city does not readily accept the newcomers they move into shantytowns on the outskirts of the city. One half of Mexico City's population of thirty million lives in cardboard and tin shacks hastily constructed along with pieces of plastic and packing crates. On the outskirts of Manila in the Philippines three thousand people live on and in a garbage dump.

These shantytowns or slums have no sewage, no trash pick-up, no running water, no electricity and no police protection. People here live in fear of the day when the city decides it no longer can tolerate them When these people do get jobs they are at low wages and often hazardous to health. People who get these jobs find themselves trapped. They make enough money to live on and they can't risk going back to their villages where conditions are even worse.

A nation can alleviate migration to the city from rural areas by offering social services such as schools and immunization services in rural areas. Building new factories in rural settings will not only employ rural people but decrease pollution in the city. Creating working committees in rural areas who can determine the problems and their correction would be a move in the right direction. This would help eliminate the fear of the central authorities, prevalent in most rural areas.

The definition of a city has expanded from just that of its political boundaries to that of metropolitan area where everyone on the outskirts of the city are dependent upon the city for services and employment. Some geographers argue that the area from Boston to Washington D.C. is one city separated by green belts. This continuation of urbanization has been given names such as Megalopolis and Bosnywash.

Future Considerations

The United States does not have an official population policy. Should a nation have such a policy? One can see official posters in China advocating one child and in India two children. Even though we do not have an official policy we affect the population conditions in other countries by giving foreign aid, food relief and settlement of refugees.

We have legislation in the United States which, to some extent, promotes certain policies affecting population. For instance, we give tax exemptions for dependents of the taxpayer. This encourages larger families.

We subsidize construction of roads, sewer lines, water lines and airports which encourages excess city populations to move to the suburbs. This in turn encourages the destruction of forests, farm lands and wetlands.

One of the controversial issues in the United States is the lack of control over our borders. Many groups have taken up this cause. Two of the more vocal are *Zero Population Growth* and *Population Environment Balance.* both based in Washington D.C. Their philosophy is that although regions of the U.S. have wide open space it is not possible to locate people there because the carrying capacity of the land has already been reached and in most cases exceeded.

The best way to get the land-people ratio stabilized is to limit immigration. Some other groups point to this seemingly idle land and encourage more immigration but when immigrants are permitted to enter the country most of them settle in the urban areas of California, Florida, New York and Texas, already our most populous states.

During the presidency of Jimmy Carter there was an agreement with the Cuban government that we would take their citizens who wished to come to the United States. We were overwhelmed with them. Premier Castro pulled a fast one and emptied his prisons and other institutions, putting the people on the boats to the United States. Our immigration service documented them and managed to find places for them throughout the country. In less than a year all of these people made their way to Florida and settled there. The criminals were especially dangerous and many of them ended up in U.S. prisons.

Another area of concern is our involvement in the population of foreign countries. Many people, including some influential legislators believe that when we send help in the form of medicines and food to needy countries, it only serves to increase their populations and these countries seem to have no sense of responsibility.

Years ago, an editorial in *Bioscience, Feb.* 1969 stated this philosophy when it said "We give food to the malnourished populations of the world that cannot or will not take very substantial measures to control their own reproductive rates" and this is "inhuman, immoral and irresponsible."

Regardless of your feelings about the subject, in the time it took to read the last ten pages, the population of the world increased by 6,300 people.

Top ten supercities of the world and their populations. If each person requires one and a half pounds of food per day and 18 gallons of water, how much of each commodity would have to be delivered to a city of 100,000 people? The cities listed below have millions of people in their service areas. You do the math.
Figures are in millions. Year 2012
Tokyo.34.9 (2) Jakarta 21.9 (3) New York 21.0 (4) Manila 19.8
(5) Mumbai (Bombay) 20.5 (6) Sao Paulo 20.6
(7) Mexico City 20.4   (8) Delhi 18.5 (9) Osaka 18.1 (10) Cairo 17.1

7. FOREST RESOURCES

We have only to look about us to see that we live in a world of wood. The people of the United States use more wood per capita than any other nation. Wood provides us with more than ten thousand different products. Indeed, it can be argued that humans are products of the forest.

The Value of Forests

Forests provide us with:
1. Fuel - More people in the world use wood for fuel than any other energy source. As the fossil fuel reserves decrease there will be a greater demand for wood as fuel. Unfortunately, some areas such as Haiti and sub-Sahara Africa have been completely denuded of forests and people living there have to rely on grass, dried manure and brush to cook their one meal a day. In places such as India, the use of dried manure for fuel deprives the soil of much needed fertilizer. This conflict of use damages the soil and further adds to the poverty of that area.
2. Wildlife Habitat - Forests contain more wildlife per acre than any other ecosystem. Forests bordering the Tundra are refuge areas for

caribou. The Tropical Rainforest contains more than half the species of vegetation and animals found on earth. Unfortunately the rainforest is being cut over at the rate of 60 acres a minute to make room for cattle ranching and cropland.

3. Flood Control - There is virtually no runoff of water from a healthy forest area. This cuts down on soil erosion as well as flooding. Denuded, gullies and overgrazed areas can be rehabilitated by planting trees.

4. Building and other Materials Almost every house in the United States contains wood as the main structural component. It does not seem likely that any other material will soon be used to replace wood in house construction. Forest products also include paper, rayon and cellulose.

5. Recreation - Our national and state forests provide a wealth of recreation. The Forest Service under the U. S. Department of Agriculture manages 156 forest districts. This amounts to over 191 million acres of trees. The national forests provide great areas for camping, fishing, hunting, hiking and photography.

6. Scientific Information - Thousands of scientific studies are conducted annually in our forest systems. These encompass such areas as entomology, natural history, game management, ornithology, and ecology as well as forestry. Methods of improving forests and forest management are always being sought.

Major Forest Regions of the United States

If the land is left untended in a wet climate the eventual result is a climax forest. For instance, if a farmer in New York State abandoned his fields and no one bothered them for ninety years and the farmer returned he would find a forest dominated by beech, maple and hemlock. This is the climax forest of that region.

The evolution of a bare field into forest progresses from (a) grass and low plants to (b) shrubs to (c) saplings to (d) forests of sun-loving species, to (e) forest of shade tolerant species. In each of these successive stages there are animals which prefer them for habitat. Each stage has its own plant and animal species.

In the United States our basic climax forest regions are:

(1) Northern Evergreens - This area stretches from Minnesota east to Maine and from southeastern Canada south to central Pennsylvania and along the Great Lakes. It is marked by beech, maple and hemlock in the southern portion and white and red pine in the north as well as spruce and fir further north.

(2) Central Hardwoods - The hardwoods, mainly of oak and and maple, inhabit the area south of the northern evergreens and continue to the northern portions of Louisiana, Mississippi, Alabama and Georgia. This is the center of our hardwood products industry. Furniture is manufactured in Ashville, North Carolina and hardwood flooring in central Tennessee.

(3) Southern Pine -This is the coniferous region of the Deep South and the home of longleaf and loblolly pine. It forms in the north along the coasts of North Carolina and covers most of Louisiana, Mississippi, Alabama and Georgia, as well as northern Florida. This was the traditional home of naval stores when sailing ships looked for sources of tar, pitch and turpentine. The straight pine trees were also sought for the main ship masts. Pine is still a major wood for home construction.

(4) Northwest Fir -The Pacific Northwest with its huge temperate rainforest has been the scene of much controversy between environmentalists and loggers. The logging industry had become accustomed to getting access to public lands at unbelievably low cost. They were cutting off the old growth forests with little thought to the future when environmentalists used the Endangered Species Act to inhibit their voracious appetites for inexpensive lumber. The large spruce and Douglas Fir trees have been momentarily saved for future generations.

The Northwest Fir region encompasses northwestern Washington and Oregon. Some Douglas fir trees are more than a thousand years old and over 200 feet high. They rival the giant redwoods which grow to their south.

5. Mixed Evergreens - South of the northwest fir and spruce grow the giant redwoods of southern Oregon and northern California. California is also home to the sugar and ponderosa pine, both valuable species of trees for the lumber industry.

6. Rocky Mountain - These are large patches of forest isolated along the slopes of the Rocky Mountains just below the snow and alpine meadows. Ponderosa, white and sugar pine are dominant here. Unfortunately, cattle and sheep are allowed to graze on much of these lands and regeneration of the forests is a problem.

Harvesting the Forests

The first European settlers to come to America viewed the forests as something to be overcome and conquered. They never gave much thought to conservation and **sustained yield.** Today we realize that we are very dependent upon forests and forest products and need to preserve forests not only for our own uses but for those of future generations.

The first Federal Forest Reserves Act of 1891 set aside timber reserves in the Yellowstone area. The act authorized the president to set aside more federal lands to ensure future timber supplies and protect water resources.

The Multiple Use and Sustained Yield Act of 1960 was an attempt to put the U.S. Forest Service in a position to govern the forests without undue influence of the logging industry. Eventually, the multiple use of forests for flood control, erosion control, preservation and recreation took a back seat to logging.

During the 1960s, 70s and 80s the Forest Service permitted unprecedented logging. Not only were trees sold to loggers at ridiculously low prices but the forest service lost money by building roads into wilderness areas for the benefit of loggers. The only national forest to show a profit during those times was the Allegheny National Forest in Pennsylvania. The country lost millions of dollars.

There are three basic methods of cutting forests - clear cutting, strip cutting and selective cutting. The type of harvest depends on many factors including financial profit as well as conservation.

Clear cutting is the standard logging practice in Maine and the Pacific Northwest. An area is cut of all timber and brush and the cut over area can be left to regenerate trees by wind blowing seeds from those border trees left standing. In some instances seeding is done by helicopter or by hand scattering. However, most forests are replanted by hand with two year old seedlings.

A clear cut area is not pleasant to see but it is an effective harvesting method. If clear cutting is done on a rotation basis then a sustained yield can be maintained. This method also permits quick harvesting of a large number of trees and for a specific purpose.

Clear cutting encourages the growth of shade intolerant trees such as pine, walnut, fir and aspen.

Clear cutting accelerates water runoff therefore encourages flooding. With an increase in water runoff there is an increase in soil erosion. Trees on the edges of clear cuts are more likely to be blown over by winds since they no longer have the protection of surrounding trees. It also limits the carrying capacity of wildlife. And, the brush and slash accompanying clear cutting is always a fire hazard.

However, clear cutting is the only method of accelerating the growth of shade intolerant species of trees. Once brush starts to grow it becomes home to deer, rabbits, grouse, elk. moose, and many songbirds. Clear cutting also can be used to remove a small portion of a forest infected with diseases or pests.

Strip cutting is a variation of clear cutting. Strip cutting involves only cutting a path of about a hundred yards through the forest leaving stands of about a hundred yards on each side. This gives the benefit of growing shade intolerant trees and fire breaks in the cut over strips. This also minimizes the loss of water and soil during heavy rains.

Selective cutting is a search and cut method of harvesting. Loggers go into a forest and search out the tree species they desire. Under these conditions trees may be harvested and a forest of uneven age is maintained. This preserves the forest ecosystem where the clear cut method does not. This system, although more expensive than clear cutting, results in a high rate of natural reseeding.

The big disadvantage of selective cutting is that the best types of timber are harvested and the less desirable types are left behind. Unless the forest is properly managed this could be a serious problem. Another disadvantage of selective cutting is that it is not useful for regenerating shade intolerant species such as pine and fir which are in big demand for lumber and plywood.

Selective cutting can be made on a rotation basis and a sustained yield can be accomplished. It also reduces the fire hazard to forests because there is little slash left behind. Selective cutting also inhibits monoculture.

Monoculture

The U.S. Department of Agriculture has encouraged farmers and large land owners to begin tree farms. These usually result in monoculture or the growing of one species of tree. The advantage to monoculture is that a particular species of tree can be grown for a specific purpose such as pulpwood or furniture veneer.

Monoculture is an efficient method of producing large volumes of timber. Since the trees are all the same age they can be harvested with a minimum of expense. It also is less expensive in the use of fertilizers and pesticides than mixed cultures. And, it makes the growing of shade intolerant species possible.

Monoculture does not encourage a forest ecosystem to develop. Once the ecosystem becomes established it is cut over and a new rotation begun. There is no development of litter on the forest floor. Monoculture fosters the overuse of fertilizers and pesticides as a solution to many growth problems.

Monoculture is more susceptible to diseases and pests. An insect infestation is a tragedy since it is easy to move from one tree to another in the monoculture tree stand whereas a gypsy moth which prefers oak trees might have a hard time finding the next one in a mixed forest.

Forest Problems

The major forest problems include insects, fungi, fires, flooding, grazing animals, storms and overcutting. As conservationists we can do something about all of these problems except storms.

Among the insects that have been especially destructive to American forests in the past are gypsy moths, bark beetles, budworms, sawflies, bagworms, tent caterpillars and weevils. The gypsy moths prefer oaks. They have slowly made their way westward from New England where they escaped from an experiment to produce silk.

Bark beetles eat the cambium layer of the bark and girdle trees. They have also been responsible for the spread of various fungi such as that which causes the Dutch Elm Disease.

Budworms kill the terminal buds of trees as well as defoliating them. Bagworms strip spruce trees of their needles and weevils eat the sprouting leaders of pine and spruce trees causing them to grow crooked. Tent caterpillars build huge unsightly webs in trees such as wild black cherry. Their larva move out from the protective nests and devour the leaves. They return to the nests for protection and to pupate and produce the moths which in turn lay more eggs.

Probably the most destructive of the forest pests are the fungi. These produce such diseases as rusts, wilts and blights. Particularly bad in pine forests are the white pine wilt, brown spot, western red rot and root rot. Oak forests are plagued with oak wilt. Elm trees get the Dutch Elm Disease.

Insects are controlled by spraying insecticides, burning infested areas and in some cases hand removal. Fungi are controlled by spraying, pruning, burning, placing mechanical barriers and premature harvesting of trees. There is also some progress in pH soil research and the incidence of fungi infections. Research also indicates that planting certain species of trees in close proximity encourages certain types of fungi and discourages some species.

Fire And The Forest

About three hundred fires break out in the American forests every day. Most of these are quickly brought under control. About ten percent of the fires are caused by lightning and ninety percent by humans. Most of the human caused fires are not accidental. Often they are set by firebugs, people who get a thrill out of seeing fires. Many are also set by people seeking revenge for real or imagined slights by the forest owners.

A controlled burn is fire set purposely by foresters. These have low flames and are constantly monitored as they spread along the forest floor. The purpose of these controlled burns is usually to get rid of underbrush which may cause a more serious fire later.

Controlled burns are also used to eliminate competition of plants with trees. This technique is used extensively in the Southern Pine region to prepare the soil as a seedbed for the next generation of trees. It also stimulates the activity of soil bacteria which is essential for good tree growth.

The National Park Service has a policy to let a natural fire burn itself out even if it means the loss of trees. The philosophy here relies on the fact that nature has created fires over the centuries in order to establish an ecosystem relying on burnt over woodlands. This policy was temporarily abandoned in the great Yellowstone Park fire of 1988 when fires threatened human habitation.

Forest fires are of three types - soil fires, ground fires and crown fires. Soil fires smolder at root level. These may burn for years before they can be controlled. It relies on heavy dry humus, moss and other almost peat conditions.

The ground fires stick to the underbrush and are low but extensive burns. Even though ground fires are considered beneficial to commercial monoculture forests they can be detrimental to a mixed forest ecosystem.

The crown fires are the most destructive as the flames leap form tree top to tree top. Deer and other creatures of the forest run for their lives. With a good wind these fires may take months to get under control. Their damage runs in the millions of dollars. Besides the Yellowstone Fire, in 1988, over 18,000 acres of forest burned in Colorado, another 17,000 acres burned in South Dakota and 15 000 acres burned in Alaska were destroyed in that same year..

Forests can be protected to some extent by cutting wide swaths through them to serve as fire lanes. Unless there is an exceptionally strong wind the fire is stopped at the lane.

The Forest Service uses spotters on high towers to watch for fires. When an obvious fire is identified the spotters call in the azimuth to a central location. By plotting the various azimuths on a map the headquarters can identify the exact location of the fire.

Once the fire is located a photography team is dispatched to the spot in helicopters and video tapes are made of the blaze. This might appear to be wasting time but the nature of the fire determines what kind of equipment is necessary to fight it. If the fire is severe, then smoke jumpers are sent in to start controlling it with techniques such as back burning. This is burning material back into the fire and when the two flames meet the fire cannot progress any further. Helicopters are also used to dump water and flame retardant chemicals on the fire.

Conservation Of The Forest

The best conservation method for preserving the forest is not to cut it. However, that is not feasible. So we have to make adjustments in our forest practices and use of wood products. We can use wood substitutes such as abundant aluminum for scarce hard furniture woods. Aluminum and steel can be used in house construction. Steel barns and small industrial buildings are an innovation that is catching on in America. We can recycle paper products to save trees as well as landfill space. Presently about half of all landfill material is made of paper.

There is a need to upgrade eastern forests of the United States. Much of the eastern forest is in the hands of private citizens and most eastern states have agriculture departments which work in close harmony with woodlot owners. They offer seminars in woodlot management and disease and pest control.

Countries such as Canada, Sweden, Norway and Finland have small populations and extensive forests which they are willing to export at prices lower than those of the United States.

One of the big world problems is the destruction of the tropical rainforest. This "green lung" of the earth is being cut at the rate of 90 acres a minute. The rainforest has more than fifty percent of all plant and animal species left on earth. The rainforest is being cut mostly to furnish specialty woods for the furniture industry and to clear land to raise hamburger cattle which is exported to the fast food restaurants of the United States and Europe.

If we lose the rainforest gene pool we will have lost one of the great riches of the earth. More than ninety percent of all prescription drugs have been developed or created from plants and most of them have been from rainforest plants. We don't know how many more life-saving drugs can be developed from rainforest plants but we do know there are unlimited possibilities there.

Controversy In National Forest Policy

Each year, enough timber is cut on America's national forests to build about one million houses, nearly twenty percent of the nation's timber harvest. Millions of head of livestock graze on the forest floor. Half of the nation's big game animals and cold water fish live there.

Environmentalists, ranchers, motorcyclists and timber companies are engaged in a fierce and emotional debate of the future of the forests. At the center is the U. S. Forest Service, the federal agency that will determine the future of the forests and the resources they contain.

The Forest Service is the largest and oldest of the public land agencies. The agency's holdings constitute a land empire larger than many countries of the world. There are 156 National Forests encompassing 191 million acres. This is equal to the total area of Illinois, Iowa, Michigan, Wisconsin and Minnesota.

Controversy erupted in the 1970s when the environmental movement challenged the growing demand by timber companies to cut more trees on public lands. This controversy became intense in the Pacific Northwest, in the "old-growth" forests. This area contains ancient Douglas fir and hemlock stands, some trees more than a thousand years old.

In assessing the controversy, Jay D. Hair, president of the National Wildlife Federation (NWF) said "Old growth is attractive to timber companies because it represents profits. The trees are large, they produce high quality timber and they are relatively accessible. In short, the economics of cutting old growth are highly favorable".

Almost all of the old growth on private land in Washington, Oregon and northern California has been cut. What remains is on public land, primarily in the national forests. If cutting at former rates continues in these forests they will be gone in by the middle of the twenty first century.

Unfortunately the Forest Service has been selling off trees to timber interests at huge losses each year. The Forest Service losses amounted to an average $343 million a year over eleven years starting in 1993. Taxpayers subsidized these losses.

It takes about one hundred years for a Lodgepole Pine to reach 65 feet. Under present regulations a timber company can buy the right to take this tree for a dollar. Also, consider the fact that the U. S. Government also builds free roads to give loggers access to these trees. In 1995, the Forest Service constructed 580,000 miles of roads through our National Forests, more than enough to reach to the moon and back. The interstate highway system in 1995 was 40,000 miles.

The Forest Service is bound by law to manage the forests for all their values - this includes preserving wildlife, protecting watersheds and providing recreation. Timber harvesting is just one aspect of the obligations of forest management, yet it dominates federal expenditures in the national forests. Many organizations such as the Sierra Club, the National Wildlife Federation and The Wilderness Society monitor the situation in our national forests. However, they do not have the financial resources to match those of the timber companies and unfortunately our political system is heavily influenced by contributions to political campaigns. This does not bode well for our national forests, one of our greatest natural resources.

Vital Statistic: Land Owned by U.S. Government % of total land in selected states: Nevada 82.3, Alaska 67.8, Utah 63.8, Idaho 62.6, California 60.9, Wyoming 48.8, Oregon 48.2, Arizona 43.3, Colorado 34.1, New Mexico 33.1, Washington 29.0, Montana 27.7

8. RANGELAND RESOURCES

Rangelands were the grazing grounds for wild animals long before there were people. These are areas of low vegetation sometimes called grasslands although grass is only one component of them. Besides grasses the rangelands have reeds and rushes, forbs and brush. Forbs are low succulent plants such as dandelions. Reeds and rushes are usually found near wet spots and these have round stems.

People of the world raise about ten billion animals on rangelands. Most of these are ruminants, animals that have cloven hooves and chew cud. Most of these are cattle, sheep and goats.

There are many different classifications of rangelands in the world. The large areas are designated as steppes, tundra and savanna. There are sub-classifications within these large ecosystems.

When European settlers first came to North America they crossed the Mississippi River and headed west. Here they found wide open spaces. Here were bison grazing in unbelievable numbers. The principle grasses at that time were Big Bluestem, Bluegrass and Buffalo Grass. Other grasses of importance to the bison were Green Needlegrass, Canada Wild Rye and Sideoats Grama. These grasses along with the forbs supported millions of animals and the native people that depended upon them for their livelihood.

In a very short period of time, from roughly 1840 to 1890 the bison were diminished to a few hundred animals. Reports state that the last bison in the wild was killed in 1894.

This was only the first of many animals that were to be depleted by the European pioneers. However, the story of the bison has a happy note. From the roughly 200 animals found in wayside zoos and a few kept by settlers as curiosities the federal government along with some concerned citizens were able to bring the bison herd back to over 200,000 million animals in the next fifty years.

When World War I was in full swing around 1917 there was a move on to raise more beef cattle and western lands under federal control were opened to full scale ranching. After the war, ranching continued and the western rangelands deteriorated since the people who used them did not take care of them and the federal government was lax in its administration of these lands.

Soon, western lands were put to the plow and the disastrous droughts of the 1930s created the famous "dust bowl" which literally blew the land away.

Conservationists were able to persuade Congress to pass the Taylor Grazing Control Act in 1934 which only affected federal lands. It didn't help much but at least the federal government was getting more involved in preserving one of our great national treasures.

The western rangelands continued to deteriorate which resulted in the creation of the Bureau of Land Management (BLM) in 1959. This organization was placed under the Department of Interior and was given the assignment of rehabilitating the rangelands. In their first inventory of lands only eighteen percent was found to be in good condition.

The BLM started a program of planting windbreaks or shelterbelts where trees were planted to cut down on wind erosion and evaporation. They also started a program of deferred grazing and allotment of cattle and sheep on ranges. A program of reseeding the desirable grasses mentioned earlier was also begun.

Unfortunately, the BLM assumed its mandate was to create good rangeland for grazing cattle and sheep and it worked toward that end. Its other mandates of preservation, conservation and recreation were pretty much ignored. The land continued to be abused and only with the election of President Bill Clinton and his appointment of Bruce Babbit as Secretary of Interior did the influence of cattle barons on Congress begin to decline.

Range Management Techniques

Management of rangelands begins with an inventory of its vegetation. Plants are classified as decreasers, increasers or invaders. Decreasers are the desirable plants such as Big Bluestem that decrease when the land begins to become overgrazed. Increasers are the less desirable plants which begin to increase with overgrazing. These include Kentucky Bluegrass which is not desirable in the western variety. The invaders are very undesirable species such as yarrow, cactus, ragweed and thistle.

Range managers also note the amount of mulch on the ground. This is the dead plant material that covers the ground under the growing vegetation. This has implications for the future organic content of the soil.

The condition of soil is also noted. A thick spongy layer of erosion resistant sod is excellent. Sometimes the pH of the soil is taken in order to determine specific needs. However, most rangeland managers can look at the vegetation and tell whether it needs fertilizing or liming. Plants such as sorrel indicate a highly acid soil. Other plants also indicate a condition of soil.

Once the condition of the range is ascertained the carrying capacity of each section can be determined. This is the ability of a particular section to graze a certain number of animals without extensive deterioration. The carrying capacity is diminished when there are wild species competing with domestic species for the range. Presently, the competing wild species include jack rabbits, elk, deer, antelope, mice, prairie dogs, rats and even grasshoppers and crickets.

Grass turns solar energy into chlorophyll and green plants. The grazing animals get their energy by eating the plants and we get our energy by eating the grazing animals. Therefore, we are a product of solar energy.

The tips of grass can be eaten without affecting its growth since the plant grows from the roots and sends up a continuous stream of blades of grass. The grass can be continually grazed as long as the grazing animals do not get too close to the sod. When they begin to eat into the sod (overgraze) they destroy the metabolic reserve of the plant. This part of the plant is necessary for plant survival since it is where the plant sends energy down to preserve and enhance the roots.

Grasses may have root systems extending down to six feet and much of the plant energy is stored in them. This enables the plant to survive long periods of drought and brief periods of overgrazing. By eating the metabolic reserve the grazing animal kills the plant.

In order to keep cattle in a certain section of range the managers or the cattlemen control the water and salt supply. If there is a stream in the section then it is difficult to get cattle to move to another section. But cattle need salt and by moving salt blocks to the new section cattle will eventually follow.

If the range is overrun with undesirable plants such as mesquite the area may be sprayed with a herbicide which will kill all shrubs. Once the shrubs are killed and dry they can be burned. This can also be done with prickly pear cactus, another plant pest of our western rangelands.

After burning, the land can be seeded with the desired grasses. Seeding used to be by hand and broadcasting, that is just throwing it around in a specific pattern. This resulted in seed loss to birds and mice. The exposed seeds were often killed by cold winter. Today, seeding is completed with machinery which punches seeds into the ground.

In African countries grasshopper locusts as they are called, get about one third of the grain crop. They also get a large portion of our western range plants. Grasshoppers multiply best in poor rangelands and if the rangeland is healthy the number of grasshoppers is severely diminished. It is better to have this natural control of insects than to blanket the land with pesticides.

Sometimes there are more jack rabbits in numbers than there are cattle on the range. A large number of jack rabbits encourage increaser and invader plants. So again, maintaining a healthy range involves control of these pests.

Cattle and sheep are also affected by the same predators that cut into the wild antelope and rabbit populations. These include wolves, mountain lions, bobcats and even bear. The worst of these is the coyote

Today around ninety thousand coyotes are destroyed annually on our federal western rangelands. Unfortunately, the ranchers want widespread destruction of coyotes but reason tells us that the coyote hunts should be only in those regions where coyotes are a menace and actually killing lambs.

Coyotes, of course, do not attack cattle. Putting cattle in with the sheep guarantees that coyotes will leave the sheep alone. Coyotes keep mice, rats and jack rabbits from overpopulation. They do more good than harm. By destroying the grass eaters, the coyote expands the range for domestic animals. The value of the added range outweighs the value of the few lambs that are taken by coyotes each year.

Major Problems of Rangelands

Our southwestern rangelands have become endangered. What was once healthy range in southern California, Nevada, Arizona, New Mexico and Texas have become deserts. Overgrazing coupled with drought have caused the land to enter into a process of desertification. This is the conversion of rangeland to desert conditions. It is extremely bad in Africa where the Sahara Desert is advancing at the rate of ten miles per year. Some progress has been made in Africa in abating this advance.

Desertification can also occur when rangeland has been irrigated over a period of years. A build-up of mineral salts forms on the surface. This necessitates plowing deeper and deeper to mix the salts with soil and eventually ends up with a sterile soil. This salinization of soil can also occur when salt lakes or ocean water infiltrates the ground water system under the rangeland.

The best solution to desertification and salinization is to take the rangeland out of production. However, this is difficult to do in places such as Africa where starvation is everybody's neighbor.

One of the solutions to rangeland deterioration might be to increase the use of native wild animals on the ranges instead of the usual cattle and sheep. Some experiments with bison, yaks, antelopes, wildebeeste, llamas and water buffalo have been successful. Reindeer and musk ox "ranching" have been successful in the cold lands of the tundra.

Rangelands make up fifty six percent of the land of North America. The cattle and sheep industry annually accounts for just under ten percent of our Gross National Product. So, it is important that we conserve and make wise use of this valuable resource.

We should be concerned about rangelands in other sections of the world since loss of potential food producing soils will affect everyone. Almost half the earth's usable rangeland is used to graze livestock. The other half is too dry or too cold for conventional ranching.

Many countries with grazing lands are too poor to curb their use. In more affluent countries such as Canada, United States, Argentina and Australia we can afford the luxury of delayed grazing and other conservation schemes.

Good rangeland management involves limiting the number of animals on a range to coincide with its carrying capacity and excluding livestock from sensitive riparian areas. In the United States there is a need for complete revision of the use of federally owned rangelands. Perhaps, we should eliminate subsidized ranching on our public lands and let the livestock industry meet free market competition.

We like to think of timbermen, miners and ranchers as rugged individuals who make their living by he-man enterprises. The truth is, almost all the timbermen, miners and ranchers are subsidized by the federal government and they are as dependent on the government as welfare recipients.

Watersheds on Rangelands

Riparian zones are strips and patches of vegetation that surround streams and wet places. Riparian zones act as green ribbons of life that bind the landscape into one ecosystem. If you change the riparian zone, you change the ecosystem.

Riparian zones make up about five percent of our rangelands but they often govern the activities on the other 95 percent. For instance, in Oregon, livestock get 81 percent of their forage from riparian zones which make up only two percent of the area.

Riparian zone ecosystems contain groups of plants, animals, soil and water that comprise the biotic communities. These zones are found not only along streams but around seeps, springs and lakes. In arid regions, they are often the only place that is green.

Riparian areas provide food and shelter for rodents and small birds as well as eagles, hawks and owls that hunt them for food. Riparian zones store water, filter it, trap sediment and reduce floods. The zones act as pipelines that carry clean water downstream for use in cities and irrigation of crops.

Riparian areas which are critical to a healthy ecosystem attract livestock. If not properly managed, livestock can damage riparian vegetation beyond its ability to recover and destroy stream-bank structures.

Livestock grazing in riparian areas often results in higher stream temperatures, excessive sediment in the stream channel, high coliform bacteria, stream channel widening which reduces water depth, change or elimination of vegetation, lowering of the water table and gradual stream channel trenching or braiding.

Beaver, waterfowl and moose may disappear from an entire drainage area once riparian shrubs die. When habitat for rodents and small birds disappears, predators leave or hunt other prey.

In most federal rangeland riparian areas, managers already know ways to reduce or eliminate livestock and other impacts, such as road construction, logging or off-road vehicles. This knowledge can't be evenly applied until range managers have time to work with livestock owners and other public interests to develop or revise grazing management plans.

Improving riparian management often requires managing livestock in radically new ways. Changing from mid-summer to late winter or fall grazing may be the solution. Or in some cases, spring grazing of riparian areas and the removal of cattle before the hot dry part of summer will allow reparian vegetation to thrive. Unless people, who own and manage the livestock, are committed to the changes needed for riparian protection and improvement, the changes and improvements will come slowly.

The Bureau of Land Management

The BLM administers what remains of the nations once vast land holdings, the public domain. This once stretched from the Appalachian Mountains to the Pacific Ocean. Of the 1.8 billion acres of public land originally acquired by the United States, two-thirds went to individuals, industries and the states. Of that remaining, much was set aside for national forests, wildlife refuges, national parks and monuments and other public purposes. This left BLM to manage about 272 million acres, about one-eighth of the national land surface.

The BLM also manages the mineral estate under 572 million acres, 300 million acres of which are administered or owned by other agencies or private interests. Most of the lands are located in eleven western states, including Alaska, although small parcels are scattered across the eastern United States.

Multiple Use Mandate

Recreation: BLM manages a full range of recreation activities including the Wild and Scenic River System, National Trails and four million acres of lakes and reservoirs used for fishing and other water activities.

Forestry: There are over 90 million acres of forested land managed by the BLM. Most of this is in Alaska but with at least 26 million acres in the lower 48 states.

Wilderness: BLM manages 25 wilderness areas in eight states. There are also about 900 wilderness study areas covering more than 25 million acres under BLM supervision.

Range: BLM manages livestock grazing on 1,265 million acres of public lands. In 2005, almost 17,000 ranchers grazed livestock on public lands. About 90 percent of these ranchers had less than 500 head operations.

Cultural Resources: There are over 150,000 identifiable cultural resources on public lands. There are 350 archaeological and historic properties entered in the National Register of Historic Places and an additional 1,200 considered to have significant values. These range from campsites of the hemisphere's earliest human inhabitants to physical reminders of the historic settling of the West.

Wildlife: BLM manages wildlife habitat for more than 3,000 species, including 140 threatened or endangered plant and animal species. It manages habitat for one out of every five big game animals in the United States, including caribou, brown and grizzly bears, desert bighorn sheep, moose, mule deer and antelope.

Wild Horses and Burros: There are more than 43,000 wild horses and burros on public lands. Any citizen with good intent can apply to the BLM for a free horse or burro. The citizen must pay for immunization shots and transportation.

Lands: The BLM issues leases, rights-of-way and use permits for a wide variety of uses of public lands and parks. This includes such things as power line transmission, petroleum product collection, motion picture filming and recreational events.

One of the controversial aspects of BLM management is the legal issuing of mineral rights to developers. The original laws passed in 1875 are still in effect. This allows developers of oil, gas, geothermal, coal, radioactive materials, building materials and metal ores to obtain the right to exploit these for two dollars and seventy five cents an acre. Most of these leases are worth millions of dollars and the system is simply taxpayers subsidizing industries which could well afford to pay the current value for these leases.

Public lands provide about fifty percent of the nation's potash, forty-five percent of its sodium compounds and seventy percent of its lead. Public lands in the West are supporting a new gold rush. Fourteen of the top twenty five producing U. S. gold mines are on public lands. The rights to these were purchased at unbelievably low prices.

Federally owned rangelands account for 55% of the total rangelands in the United States. These lands are no longer wild and must be managed with an eye to preservation and future use. Can we continue to manage these once wild lands for domestic livestock and eliminate the native animals which once roamed freely on them? Are we now the caretakers of all land and animals and do we have an obligation to those animals we have displaced?

9. WILDLIFE RESOURCES

Human life and human experiences have evolved with wild plants and animals. We owe our very existence to these wild resources. In our distant past and in our present we have taken many of these wild plants and animals and domesticated them. That is, we have taken them from the wild habitat and created habitats for them in order for them to be more accessible to us.

Generally, when we use the term wildlife, we are referring to the animal world. However, we must always keep in mind that we cannot separate the wild animals from the wild plants which are necessary for their survival, and ultimately for our own survival.

Wildlife resources afford humans many benefits. These benefits can be classified as (1) economic (2) scientific (3) aesthetic, and (4) recreational. Each has its own value in making life richer and more comfortable for us.

Our economy is spurred on by the benefits from anglers, hunters, campers and bird watchers to mention only a few of the groups. These people spend millions of dollars pursuing their hobbies, which enhances the general economy.

For instance, there are over seventeen million hunters in the United States who buy licenses each year. The states of Texas and Pennsylvania each sell over a million hunting licenses annually. Pennsylvania's economy benefits directly from the purchase of the licenses and other fees to the tune of 500 million dollars annually.

These hunters also spend money in motels, hunting lodges and restaurants each hunting season. They spend millions of dollars on guns and ammunition and hunting clothes. In the United States the hunting dog industry nets over ten million dollars each year.

The recreational uses of our wildlife and habitats include camping, hiking, trail bike riding, snowmobiling and bird watching. Birders spend over a hundred million dollars a year just visiting birding "hot spots."

The scientific study of animals has given us many insights into our own behavior and illnesses. Practically every break-through in modern medicine has been by experimentation with animals. By studying bird migration and animal hibernation we know now that many people get depressed in winter due to the lack of sunlight.

In many parts of the world, wildlife is a necessary supply of food. Even in American society we consume over ten million grouse, nine million deer, twenty million wild rabbits and two million hare each year. All of these harvested from the wild.

Extinction

The fossil record indicates that millions of animals which once made earth their home are now extinct. They and their genes have disappeared from earth. We need not look at the fossil record to see that extinction is a fact of life, we only need look to our own recorded history. Birds such as the heath hen, passenger pigeon and great auk have all become extinct during the lifetime of the human species.

We like to blame such events as extinction on a particular circumstance, but extinction is a combination of factors.

Some of these factors are:

Low Biotic Potential: Each species has a theoretical maximum of population growth if the resources available to them are unlimited and they are not hampered by such things as disease, predators and competition. If two flies mated in early March and their offspring and the offspring's offspring survived to maturity, then by the end of November the land portion of the world would be covered with flies to a depth of five feet.

This is a theoretical situation but we have many examples of real life events that have occurred in the past. A famous study of mule deer in the Kaibab National Forest of the Grand Canyon illustrates this point well. In 1906 the State of Arizona placed a bounty on the wolf, coyote and cougar. The National Forest is inside the boundaries of Arizona. Within fifteen years the predator population had been reduced to only a few hundred animals and the wolf population was wiped out completely. In the next twenty years the deer population which started out at about five thousand animals had soared to over a hundred thousand. This was well beyond the carrying capacity of the habitat. The deer ate just about everything they could reach and in the next six years the population began to crash with an annual starvation rate of sixteen thousand deer. The forest was severely damaged since a browse line emerged and no new seedlings were permitted to germinate since the deer ate these as fast as they came up. This forest has still not recovered.

Competition and Survival: There are three stages of animal populations (1) production (2) competition, and (3) survival. Each animal species is capable of producing more offspring than the habitat will support. For instance, a pair of adult oysters will produce almost two million eggs. When those two million eggs hatch the young are subjected to intense competition for food. They must fight each other for food and they must fight other species such as clams, crayfish, squid and fish for the same food. Those oysters are also food for other species. Of those two million original oysters only two will survive to maturity. Those that survive will pass their survival genes on to the next generation. As a result, we continue to get oysters better adapted to the environment. Keep in mind that you are here because your ancestors knew how to survive. They overcame tremendous obstacles in order to reproduce and have you as a final product.

Food Preference : We classify animals into several categories for easy identification. Herbivores are those animals that only eat plants. Carnivores only eat meat and omnivores eat both plants and meat. The feeding range of carnivores is larger than the range of herbivores. Carnivores are predators or scavengers and when the population of herbivores decreases so does the population of carnivores.

Some animals have such specialized diets that they are in danger of extinction. The panda eats only a few species of bamboo shoots, the koala bear eats only eucalyptus leaves and when these plants are threatened by disease or bad weather then the animals that feed on them are likewise threatened.

The Everglade Kite, a large eagle-like bird, feeds only on one species of snail. If the snail should suddenly succumb to a disease or pollution then the Everglade Kite must change its eating habits or perish.

Predators and Parasites: Predators are animals that hunt and kill other animals, which is the prey. Parasites are animals that live on or off other animals but do not necessarily kill the host animal. Cougars, coyotes and wolves are predators. These eat such animals as birds, rabbits and browsers.

The parasites of the animal world are mostly worms such as tapeworms, lungworms, heartworms and various fly larva. There are also parasites which cause diseases. These are mostly protozoans and viruses.

Technically, plants are preyed upon by browsing animals and they too have parasites which limit their populations. We must not forget that plants have similar DNA (stuff that makes up our genes and chromosomes) as animals. We are all interrelated in that we have the same basic biological molecules within us.

Life Cycles: Wildlife studies have given us some convenient methods of identifying the population patterns of animals. Most of these are based on carrying capacity, that is, the population of a species that a habitat can support on a sustainable basis.

A stable population may have its ups and downs from year to year but basically it remains at the same average number of individuals over a period of years. These populations have a good balance of food, predators, competition and reproduction.

An irruptive population is one that rises sharply. This may occur when there is a series of favorable weather conditions which increases the food supply beyond that which keeps the population in balance. Once the irruption occurs the condition will continue for a number of years until the predator or lack of food in the ecosystem brings it back to normal. In the case of the mule deer of the Kaibab the destruction of predators was largely responsible for the irruption.

A cyclic population peaks and then crashes periodically. The cycle is predictable and is related to reproduction, predators, and since this usually occurs in herbivores to overgrazing. Most people are familiar with the lemmings of the far north. Their principal predators are the fox, owl and lynx. The lemmings have a population burst every three years. This is due to factors listed previously. As the lemmings increase, an increase in fox and owls is a year behind. Due to the increased food supply they increase their populations and so the northern owls and foxes peak about every three years, just after the lemmings.

Other animals which have a cyclic population include the red-tailed hawk and the snowshoe hare. There is also a cycle which includes muskrats (prey) and mink (predator).

Habitat Destruction: As our population grows we must expand our activities to compensate for this growth. People need food, clothing and shelter as well as other necessities and non-necessities to make our life comfortable. As we engage in building cities, highways, cutting trees and farming we destroy habitat needed by wildlife. What seems to be left in America are islands of wildlife habitat called ecological islands. These are small areas surrounded by human habitation and influence. As these islands are further diminished by encroachment of humans they support less wildlife. Each species has its own critical population size. If the population falls below this critical level then it is too small for the species to survive. Also, these smaller island habitats encourage in-breeding which produces inferior offspring.

We have tried to preserve wildlife by creating a series of parks such as our National Parks and National Forests. However, these are not enough to keep many species from biotic impoverishment. Many studies have confirmed that the number of species in our national parks have been decreasing over the years. Many parks are simply too small to support major predator species such as the grizzly bear and wolves.

These deficient parks include Jasper-Banff series of parkland in western Canada and the Yellowstone complex of Wyoming. Glacier National Park has done well in protecting most of its species but the rest of the park system has lost over twenty five percent of their original numbers.

Hunting: Generally hunting is divided into commercial hunting and sport hunting. Commercial hunting produces food and other products for a large number of people. Commercial hunting has contributed to the extinction of the passenger pigeon and the Great Auk, a large bird that once lived along the North Atlantic coast. Commercial hunting has been responsible for the near demise of the bison, rhinoceros, tiger and whale. Today, commercial hunting in Sierra Leone and other west central countries of Africa are killing off the monkey populations. These are used for food. Unless some international agency interferes, the monkeys of that region will be severely diminished or gone in the next few years.

No animal that has ever been under the control of game commissions has ever become extinct. Through taxes, sport hunters have been responsible for the purchase of thousands of acres of wilderness which have been protected from urbanization and farming. These gamelands protect, not only game species, but hundreds of other non-game species as well. There are over 400 National Wildlife Refuges in the United States and these were purchased largely from the taxes and license fees of duck and waterfowl hunters.

Non-adaptive Behavior: Many species cannot adapt to changes in their environment Deer, which has a range of about one square mile, refuse to move out of that territory even as it is being cut and destroyed. Moving deer to another location has not proved to be successful. Only about four out of a hundred deer survive the move to a new location. The same is true for rabbits. Trapped rabbits which are moved to a new habitat survive less than six months and the luckier ones survive up to a year.

Protecting Our Wildlife Heritage

The Endangered Species Act was first passed in 1973. Since then, it has been modified several times. The act requires the U.S. Fish and Wildlife Service to identify species and classify them as endangered, threatened or not in danger. Endangered species are those that are in imminent danger of extinction. In the United States the latest list includes 36 mammals, 57 birds, 60 fish, 8 reptiles, 6 amphibians, 14 snails, 16 insects, 378 plants, 11 crustaceans and 50 clams. However, worldwide there are many more species threatened with extinction. Of these, there are 234 mammals and 141 birds.

A threatened species is one that is likely to become extinct if present practices involving them continue. In the United States this involves 5 mammals, 15 reptiles, 8 birds, 32 fish, 4 amphibians, 7 snails, 6 clams, 2 crustaceans, 9 insects and 8 plants. The number worldwide is not high in this category but nevertheless those few species must be protected lest they move into the endangered category.

The Endangered Species Act also requires the U. S. Fish and Wildlife Service to identity the habitat of the endangered species and once identified, prevent destruction of that habitat. This part of the law was used successfully to stop the cutting of many old growth forests of the Pacific Northwest. The spotted owl received much publicity but the argument was not over owls but over cutting the old growth forests which is part of our American heritage.

The least costly method of saving plants and animals is by setting aside large tracts of land where a population of plants and animals can continue to survive without molestation by humans. Unless the habitat is extremely large, the animals and plants are afforded only temporary protection.

A costlier but effective approach to ecological preservation is the restoration of previously damaged lands. The industrialized world is busy making deals with rainforest countries to preserve this valuable asset. The rainforest has more than half the world's species of plants and animals. However, rainforests are located in poor countries and it is difficult to tell a farmer clearing the rainforest that he can't do that and we in the richer countries would prefer he returns to his village and live in poverty.

Since the rainforest countries are deep in debt there is some hope in the process of exchanging rainforest protection for the cancellation of debt. Many holders of the debt have agreed to do this and countries such as Brazil have participated in this debt exchange for rainforest preservation.

Why Extinction Should Concern Us

Probably the biggest concern with extinction is the loss of the genetic pool. Animals such as the rhinoceros took millions of years to evolve. It has a gene pool unlike no other creature on earth. When it is gone, its particular genes are gone forever. The creation of gene banks, where DNA from different animals is preserved, is not a substitute for the animal itself

If a particular plant or animal is lost to the world we may discover that it was vital to our own survival. It may have been a link in our food chain or somehow protected us from disease.

Another aspect of preservation of species is the selfish fact that many of them have given us medicines as well as food. More than ninety percent of all drugs sold over the counter originally came from wild resources.

A drug which effectively fights childhood leukemia comes from an obscure periwinkle plant found on the Island of Madagascar. A drug, taxol. Derived from the bark of a yew tree is effective in fighting uterine cancer in women.

We don't know how many plants and animals out there have potential for making our lives more comfortable and extending our own lifespans. If they become extinct we will never know.

Factors of Wildlife Habitat

Like every other living animal on earth, wild animals need food and shelter. Shelter, which is cover, protects the animal from predators, humans and adverse weather conditions.

The habitat provides the animal with food and water. Even the desert animals must get water somehow. The birds of the desert usually fly to a waterhole and immerse themselves. They fly back to the nest and shake this water loose into the open mouths of their nestlings.

Animals can survive for many days without food but only a few days without water. A person can go at least a month without food and not suffer any ill effects. A week without water makes dehydration conditions in the body almost impossible to reverse.

Territory is an area of habitat defended by a species. Establishing territory assures the species of a supply of food, an area to mate and raise young and reduce predation. It also prevents disease from moving rapidly thorough a population.

Territory needs vary greatly from species to species. A deer might stake out a square mile but a black bear needs at least forty square miles.

It must also be kept in mind that creating habitat for one species, destroys it for another. If you create open areas for rabbits, you destroy the woodlands for squirrels.

Today, the habitats of many animals are being destroyed by urbanization, new farming practices and pollution. Acid rain has decimated many forests and caused life forms in lakes to disappear. Lead poisoning has developed in waterfowl which mistake lead shot of hunters for seeds in water. Oil pollution killed over a hundred thousand waterfowl a year in each year of the 1980s. Farming practices has filled the air with pesticides and polluted streams with fertilizers and eroded soil. We need the food but we must also insure that the production of food is carried on with preserving our environment.

Each species needs a particular habitat. We can enhance waterfowl habitat by protecting wetlands, creating openings in marshland to facilitate waterfowl movement, construct artificial ponds and swamps, and construct artificial nesting islands in large water areas. The best protection afforded waterfowl has been the establishment of the National Wildlife Refuge System.

In the United States, every state has its own game and fish commission which has been entrusted with the development of state lands for preservation of wildlife. The people who operate and manage these lands have been trained in modern scientific management techniques and we can feel comfortable that at least at this level our wildlife is in good hands. However, we must guard against developers who have money and use it to influence our legislators and who have historically tried to intrude into protected habitat. Wetlands will always be a source of conflict between those who would protect them and those who would "develop" them.

10. WETLAND RESOURCES

(The following discussion is based on publications by the U. S. Fish and Wildlife Service and the U.S. Environmental Protection Agency. It pertains only to the United States)

Wetlands include the wide variety of marshes, swamps and bogs that occur throughout the country. They range from red maple swamps and black spruce bogs in the northern states to salt marshes along the coasts, to bottomland hardwood forests in the southern states to prairie potholes in the Midwest to playa lakes and cottonwood willow riparian wetlands in the western states to the wet tundra.

Wetlands usually lie in depressions or along rivers, lakes and coastal waters where they are subject to periodic flooding. Some, however, occur on slopes where they are associated with ground water seeps. Conceptually, wetlands lie between well drained upland and permanently flooded deep waters of lakes. rivers, and coastal embayments. Recognizing this, one must determine where along this natural wetness continuum wetland ends and upland begins. Many wetlands form in distinct depressions or basins that can be readily observed. However, the wetland-upland boundary is not always that easy to identify. Wetlands may occur in almost imperceptibly shallow depressions and cover vast acreages. In the Prairie Pothole Region, wetland boundaries change over time due to varying rainfall patterns. In these situations, only a skilled wetland ecologist or other specialist can identify the wetland boundary with precision.

Wetlands were historically defined by people working in biology or hydrology. A botanical definition would focus on the plants adapted to flooding and/or saturated soil conditions, while a hydrologist's definition would emphasize the position of the water table relative to the ground surface over time. A more complete definition of wetland involves a multidisciplinary approach. The Fish and Wildlife Service has taken this approach in developing its wetland definition and classification system.

Definition of Wetlands

In developing an ecologically sound definition of wetland, it was acknowledged that there is no single, correct, indisputable, ecologically sound definition for wetlands, primarily because of the diversity of wetlands and because the demarcation between dry and wet environments lies along a continuum. Previous wetland definitions grew out of different needs for defining wetlands among various disciplines, e.g. wetland regulators, waterfowl managers, hydrologists, flood control engineers and water quality experts. The Fish and Wildlife Service specifically defines wetlands as follows:
"Wetlands are lands transitional between terrestrial and aquatic systems where the water table is usually at or near the surface or the land is covered by shallow water. Wetlands must have one or more of the following three attributes (1) at least periodically, the land supports predominantly hydrophytes; (2) the substrate is predominantly undrained hydric soil: and (3) the substrate is nonsoil

and is saturated with water or covered by shallow water at some time during the growing season of each year".

In defining wetland from an ecological standpoint, there are three key attributes: (1) hydrology - the degree of flooding or soil saturation; (2) wetland vegetation (hydrophytes), and (3) hydric soils. All areas considered wetland must have enough water at some time during the growing season to stress plants and animals not adapted for life in water or saturated soils. Most wetlands also have hydrophytes and hydric soils present. A definition similar to this was used in Federal Court in Louisiana to make a legal wetland determination. In his ruling the judge decided that the area in dispute constituted wetland according to Section 404 of the Clean Water Act because records showed that virtually most of the land was flooded every other year, the soil types were classified as wetland soils and vegetation capable of surviving and reproducing in wetlands predominated the site. The rationale for using these three key attributes now has legal precedent.

When soils are covered by water or saturated to the surface, free oxygen is usually not available to plant roots. Most plant roots must have access to free oxygen for respiration and growth. Flooding during the growing season presents problems for growth and survival of most plants. In a wetland situation, plants must be adapted to cope with these stressful conditions. If flooding occurs only in winter when the plants are dormant, there is little or no effect on them.

Permanently flooded deepwater is not included in the definition of wetland. Instead, these waterbodies, generally deeper than six feet, are defined as deepwater habitats, since water and not air is the principal medium in which dominant organisms must live.

The Fish and Wildlife Service spent four years on developing their definition of wetland. This definition is accepted as the national and international standard for identifying wetland.

Major Wetland Types of The United States

Wetlands occur in every state and due to regional differences in climate, vegetation, soil and hydrologic condition, they exist in a variety of sizes, shapes and types. Wetlands can even exist in deserts.

The Fish and Wildlife Service's classification system groups wetlands according to ecologically similar characteristics. It divides wetlands and deep water habitats into five ecological systems. These units are (1) Marine (2) Estuarine (3) Riverine (4) Lacustrine, and (5) Palustrine.

The marine system generally consists of the open water associated with coastlines. It is mostly a deep water habitat system with marine wetlands limited to intertidal areas such as beaches, rock shores and some coral reefs.

The estuarine system includes coastal wetlands like salt and brackish tidal marshes, mangrove swamps and intertidal flats, as well as deepwater bays, sounds and coastal rivers.

The riverine system is limited to freshwater river and stream channels and is mainly a deepwater habitat system.

The lacustrine system is also a deepwater dominated system but includes standing water bodies like lake reservoirs and deep ponds. The palustrine system encompasses the vast majority of the country's inland marshes, bogs and swamps and does not include any deepwater habitat.

Importance of Wetlands

Although often used by many people for hunting, trapping and fishing, wetlands were largely considered wasteland whose best use. could only be attained through reclamation projects such as drainage for agriculture or filling for industrial or residential development. Much to the contrary, wetlands in their natural state provide a wealth of values to society. Wetland benefits can be divided into three basic categories: (1) fish and wildlife values, (2) environmental quality values, and (3) socio-economic values.

Fish and Wildlife Values include fish and shellfish habitat, waterfowl and other bird habitat, furbearer and other wildlife habitat, environmental quality values, water quality maintenance, pollution filter, sediment removal, oxygen production, nutrient recycling, chemical and nutrient absorption, aquatic productivity, micro-climate regulator and world climate ozone layer.

Socio-economic Values include flood control, wave damage protection, erosion control, groundwater recharge and water supply, timber and other natural products, energy source (peat), livestock grazing, fishing and shell fishing, hunting and trapping, recreation aesthetics, education and scientific research.

Forces Changing Wetlands

Wetlands represent a dynamic natural environment that is subjected to both human and natural forces. These forces directly result in wetland gains and losses as well as affect their quality.

Human Threats include (1) Drainage for crop production, timber production and mosquito control. (2) Dredging and stream channelization for navigation channels, flood protection, coastal housing developments and reservoir maintenance. (3) Filling for dredged spoil and other solid waste disposal, roads and highways, and commercial, residential and industrial development.(4) Construction of dikes, dams, levees and seawalls for flood control, water supply, irrigation and storm protection. (5) Discharges of materials (e.g. pesticides, herbicides, other pollutants, nutrient loading from domestic sewage and agricultural runoff and sediments from dredging and filling, agricultural and other land development into waters and wetlands. (6) Mining of wetland soils for peat, coal, sand, gravel, phosphate and other materials.
Indirect Human Threats include (1) sediment diversion by dams, deep channels and other structures. (2) Hydrologic alterations by canals, spoil banks, roads and other structures. (3) subsidence due to extraction of groundwater, oil, gas, sulfur and other minerals.
Natural threats include (1) subsidence (2) natural rises of sea level (3) droughts (4) hurricanes and other storms (5) erosion (6) biotic effects, e.g. muskrats, nutria and goose eating.

Natural events influencing wetlands include rising sea level, natural succession, the hydrologic cycle, sedimentation, erosion, beaver dam construction and fire. The rise in sea level both increases and decreases wetland acreage depending on local factors.

Human actions are particularly significant in determining the fate of wetlands. Unfortunately, many human activities are destructive to wetlands, either converting them to agricultural or other lands or degrading their quality. Key human impacts include drainage for agriculture, channelization for flood control, filling in for housing, highway and sanitary landfills, dredging for navigation channels, harbors and marinas, reservoir construction, timber harvest, peat mining, oil and gas extraction, strip mining, groundwater extraction and various forms of waste disposal.

Wetland losses and degradation continue throughout the country. There are several areas where wetlands are in greatest jeopardy from a national standpoint. These areas are (1) the estuarine wetlands of the U.S. coastal zone (2) Louisiana's coastal marshes, (3) Chesapeake Bay's submerged aquatic beds, (4) South Florida's palustrine wetlands, (5) Prairie Pothole Region's emergent wetlands, (6) wetlands of Nebraska's Sandhills and Rainwater Basin, (7) forested wetlands of the Lower Mississippi Alluvial Plain, (8) North Carolina's interior swamps, and (9) riparian wetlands. Most of these regions are under intense pressure from agricultural interests.

Future of America's Wetlands

The U. S. population is growing by 1.9 million each year. More than half the population lives within 50 miles of a major coast. Pressures to develop estuarine and palustrine wetlands in coastal areas will remain intense, despite the existence of laws to protect these wetlands. As the population swells in the uplands, public managers will be greatly challenged to protect wetlands from future developments.

The population move to the sunbelt states of the Southeast and Southwest will increase urban and industrial development pressures on wetlands. There will be competition for water between agriculture and non-agriculture users. Fish and wildlife will probably lose out.

The movement of people from urban cities to suburban locations have reduced agricultural land in rural areas. These suburban counties have threatened the remaining wetlands. Since most states do not have wetland protection laws, federal regulation through the Clean Water Act is the key means to protecting these wetlands.

Increases in world population will create a demand for American farm products and this in turn will put pressure on the protection of wetlands. This demand for American grain has already led to the conversion of vast acreages of bottomland forested wetlands to cropland in the Mississippi Alluvial Plain.

American farms are experiencing a decline in returns per unit of cost and this forces farmers to increase production in order to maintain the same level of income. Since much prime farmland has been converted to non-agricultural uses there has been a tendency to make up for this by conversion of rangelands and wetlands to cropland. The use of irrigation by lowering water tables has increased the destruction of wetlands, especially in the Great Plains.

Agriculture has always played a role in degrading water quality, fish and wildlife habitat and the quality of wetlands. About 68% of all water pollution in the U.S. is caused by agriculture, with soil erosion from cropland being the single greatest contributor to stream sediment. Improved soil management practices are of utmost importance on present farmland.

Wetland protection can be accomplished by acquisition of wetlands and regulation of wetland uses. The use of tax incentives to encourage preservation of wetlands by landowners represents a potentially valuable tool in protecting wetlands. The removal of government subsidies which encourages wetland destruction would also benefit wetlands greatly.

The acquisition of wetlands for migratory birds has been especially successful in protecting them These have been acquired under The Migratory Bird Conservation Act of 1929, The Migratory Bird Hunting and Conservation Stamp Act of 1934, and the Land and Water Conservation Fund Act. Landowners can receive reimbursement through the Soil Conservation Service's Water Bank Program.

Federal funds and state governments cannot be expected to protect all of our wetlands. However, wetland regulations at the Federal and state levels are vital to preserving our wetlands and saving the public values they provide.

Property Rights vs The Common Good

In the United States we are used to thinking in terms of "individual freedom and rights". Asian cultures think in terms of the rights of the community or society taking precedent over those of an individual. In some instances, we put the welfare of society above that of the individual when we condemn property for such uses as schools, highways, hospitals or for national defense.

As our population increases there will be more demand for food, homes, roads and schools. We will be forced to make decisions .concerning which lands to preserve and will be used to supply those necessities. The problem reaches emotional heights when people who own wetlands want to destroy them and convert them to economic uses. Most people who want less government regulation on the use of their land belong to groups identified as "Wise Use" groups. These groups are particularly offended by any wetland regulations which prevent them from draining or filling in the wetlands on their "private property."

Property rights involve entitlements, privileges and limitations which define the owner's right to use a resource, in this case property. Wetlands are particularly vulnerable since many of them are in high use areas. Coastal wetlands are prime recreation and condominium properties. Prairie wetlands are in prime farm areas. Almost all wetlands are vulnerable to development.

Destruction of one acre of wetland may have little impact on a watershed but when many one acre plots are destroyed it has serious environmental implications. When we legislate to keep people from using their property, in this case wetlands, aren't we in effect condemning the property for the good of society. So, where do individual property rights end and society's resource rights begin?

U.S. Cities 100% Dependent on Ground Water include: Albuquerque, Baton Rouge, Dayton, Ft. Lauderdale, Jacksonville, Orlando, Riverside, San Antonio, St. Petersburg, and Tucson.

The per capita use of water is highest in Tacoma where citizens use 691 gallons a day. The next highest per capita use cities are Atlanta 472, Washington DC 419, Baltimore 342, Ft. Lauderdale 330 and Dayton 318.

13. ENERGY RESOURCES

Every individual needs to use fuel of some sort to carry on the basic necessities of life. Even in the tropics food must be cooked on a daily basis and in most areas of the world we must have some sort of heat to keep warm.

Fuel is just one source of energy which is defined as the ability to do work. Energy of one sort or another is necessary for an industrialized world.

There are five basic fuels used in the world. These are coal, petroleum, natural gas, wood and nuclear materials. The first three, coal, petroleum and natural gas are referred to as fossil fuels since these were created from organic matter buried millions of years ago. These are the premier sources of energy since they are comparatively inexpensive and produce a lot of energy per unit of weight. These are nonrenewable resources.

Coal

Coal was formed by ancient plants buried in the sediments of the geologic past. These were once great forests which sank under the sea and were later buried by sand, mud and gravel.

Accumulated vegetation which has not undergone the pressure and time of coal is peat. This is compressed plant matter which can be cut and transported in blocks and burns readily. Several countries of the world depend heavily upon peat as a source of home fuel. Among these are Ireland, Estonia and Lithuania.

The beginning of coal formation occurs when vegetation has undergone some form of change due to pressure upon it. Many of the volatiles, gases of the vegetation, are driven off and the carbon in the plant material is concentrated. When the concentration of carbon reaches about thirty percent the material is known as brown coal or lignite. The United States has extensive lignite deposits in the Rocky Mountain states from Arizona and New Mexico in the south to Montana and North Dakota in the north.

Lignite is a sedimentary rock. When it is subjected to more pressure and given more time to age it becomes bituminous or soft coal. This is the most desired type of coal and it is used extensively in the iron and steel industry and to produce electric power. Lignite and bituminous coal account for 56 percent of all the electric power produced in the United States. We have large bituminous coal deposits in the Appalachian Plateau, the Illinois Basin, Michigan and a large field from Iowa to Texas. The carbon content of soft coal is anywhere from 70 to 86 percent.

When the sedimentary coal is compressed and undergoes geologic metamorphism it becomes anthracite or hard coal. In anthracite the carbon content reaches 96 percent and it burns with a bright blue flame once it is ignited. Anthracite deposits are found in eastern Pennsylvania, Rhode Island and western Virginia.

Coal is obtained by strip mining and by deep surface mines. In strip mining the top layers of earth and rock are removed with the aid of power shovels. In large operations this overburden is removed with huge shovels called draglines. These can remove almost ten tons of overburden with one scoop. Of course this takes a lot of energy and unless the coal is at a shallow depth it is not economically feasible to remove the overburden with these expensive methods.

Strip mining tears up the land and causes much damage to nearby streams. The Surface Mining Control and Reclamation Act of 1977 required all strip mining companies to restore land to its original contour and to replant it after the operation. Many companies have found that it is cheaper to pay the imposed fine than to spend time reclaiming the land.

Strip mining if not restored is an ugly sight to behold. A popular blue grass song at the height of strip mining in Kentucky had various names, one was Paradise. It went "Daddy won't you take me to Muhlenberg County, along the Green River where Paradise lay – sorry my son, you're too late in asking, Mr. Peabody's coal train has hauled it away."

Shaft mining involves a large elevator shaft which transports men and materials into the ground to the level of the coal seam. In level mines the diggers operate machinery with ease. However, in a slope mine there are many dangers.

Shaft miners are ever watchful for the build-up of poisonous methane gas which can kill them. An accumulation of methane is also subject to explosion and many miners in every coal producing country have been killed by mine explosions or trapped behind the cave-ins which occur with these explosions.

Shaft miners remove much material such as slate and shale which comes with the coal. This waste material is sent to the surface in small rail cars and dumped on the land. These create eyesores and are one of the biggest ingredients in landfills.

All deep mines have water accumulation and this must be pumped out if miners are to operate in safety. This water plus the water which normally runs through coal seams is acid in nature. The streams into which they flow are devoid of life and the water is an orange-brown Once a mine is abandoned some of the water can be contained and treated as it leaves the mine but this has proven to be a difficult and costly enterprise.

The United States has the largest coal reserves in the world, 29%, followed by Russia 24% and China 11 %. At the present rate of consumption the world coal resources will last well over two hundred years. In 2011 the countries leading in coal production in descending order were China, United States, India, Australia, Indonesia, South Africa, Russia, Kazakhstan, Poland and Columbia.

Coal can be converted into a gasoline type product known as Synfuel. It can also be gasified and used in the manner of natural gas. In some deep and inaccessible deposits the coal bed can be burned and hydrogen, carbon monoxide and methane released can be captured and used. But all of these are currently more expensive than using petroleum and natural gas. But, not for long.

Petroleum

Most people refer to petroleum as oil and we shall do that also. For centuries oil was found in surface seeps and on water and used to light smoky lamps. When it was discovered that kerosene, a brightly burning product could be made from oil a search for new ways to obtain oil was underway. Edwin Drake was able to get significant amounts of oil by drilling a well through pipe casing. His successful well at Titusville, Pennsylvania in 1859 launched the oil industry as we know it today.

Oil is a liquid which is easy to transport and store. Most of it moves by pipeline on land and by ocean tanker from land to land. Heating oil is much cleaner burning than coal and there is no ash waste left for disposal.

Oil can be broken down into many products such as gasoline, kerosene, and heating oil. It is the mainstay of the plastics industry. The main use of oil is in transportation, running the vehicles that we operate and ride in or on, from snowmobiles to jet planes.

Russia, the United States and Saudi Arabia are the big producers of oil each year. Perhaps it is better to say that they are the big extractors of oil each year since the material was produced by natural processes. The countries with the largest oil reserves are Saudi

Arabia with 24% of the world total reserves, followed by Russia 14%, Iraq 9%, Kuwait 9%, Venezuela 6%, Iraq 6% and United Arab Emirates 6%. The United States which uses about one third of the world's oil supply each year has only three percent of the world reserve. At the present rate of consumption the usable oil of the world will be gone by the year 2030. Some "experts" say 2040.

The United States oil reserves are located mostly in Texas, California, Oklahoma and Louisiana. Smaller deposits are found scattered around the country.

Except for Venezuela and Mexico most of Latin America was believed to be devoid of oil. However, a large pool was discovered in Brazil. Perhaps, our estimate of world oil reserves is inaccurate. Large countries with very little known oil reserves include India, Germany, France, Pakistan, Australia and Thailand.

When it looked like the United States was going to lose its international sources of cheap oil the government began exploring the possibility of exploiting its large deposits of oil shale found in Colorado, Wyoming and Utah. This is a fine grained sedimentary rock impregnated with oil. When crushed and heated the rock gives off an oil material known as kerogen. This oil can be refined to produce gasoline and other products associated with oil. If necessary this source of oil will last us for at least thirty years. Mining and processing oil shale is obviously detrimental to the environment.

There are also deposits of tar and oil sands around the world. One of the largest deposits of oil sand is found in western Canada near the Athabaska River. Canadians are busy extracting and experimenting with these deposits. By year 2020, Canada expects its oil sands to produce half of its domestic oil supply.

In July 2010 Russia was pumping out 10 billion barrels of oil a day. Saudi Arabia was just under ten and the United States was close to 9 billion barrels a day. The other countries producing significant oil in descending order were Iran, China, Canada, Mexico, United Arabic Republic, Brazil, Kuwait, Venezuela and Iraq. There were 113 different countries producing oil in 2010.

Natural Gas

Natural gas is found in association with coal and oil deposits. It is the ideal fuel since it burns completely and is easy to transport. It is most used in industry, especially the glass industry and in home heating and cooking. It, however, is poisonous when inhaled in quantity and explosive when ignited.

At the present rate of consumption, world natural gas supplies should last about forty years. The countries with the largest natural gas reserves are　Russia 25%, Uzbekistan 15%,　Iran 14%, Qatar 5% and United Arab Emirates 4 %. The United States has slightly less than three percent of the world reserve.

Most of the Russian supply of natural gas is piped west to countries of Europe. Even though Russia has enormous gas reserves compared to the rest of the world it is found in cold and remote areas that make exploitation difficult.

In 2010 the leading countries in natural gas distribution were Russia, Iran, Qatar, United States and Canada.

Nuclear Energy

Nuclear power comes from the reaction of radioactive uranium. When a uranium 235 atom is struck by a neutron given off by another uranium 235 nuclei, the uranium atom is split into fragments. As the reaction continues other atoms are split. This is nuclear fission and it produces heat. The breakup of the atom releases its binding energy which is tremendous.

About ninety seven percent of uranium ore contains U 238 which is stable. Less than one percent of uranium ore is uranium 235 which is radioactive. In order to make nuclear fuel the U 235 is concentrated. (enrichment). This enriched fuel is inserted into steel tubes or fuel rods. The rods are about twelve feet long. bundled together according to fuel desired, usually in bundles of 100 to 300 rods.

These fuel rods are lowered into the reactor core where the nuclear reactions take place. The intensity of the chain reaction in the fuel rods is controlled by neutron absorbing control rods. These control rods in the U.S. are cadmium or boron steel. The control rods of many European reactors are made of carbon.

The control rods are raised or lowered in the reactor to control the rate of fission. By lowering the control rods completely into the reactor, fission can be completely shut down or neutralized.

The energy from the fission of the fuel in the rods heats confined water to more than the boiling point. This water is circulated past other water in separate pipes which is heated to boiling and produces steam. This steam is used to turn turbines and produce electricity.

The reactor vessel is housed in a containment building. If the reactor core is on the way to becoming overheated the containment building can be flooded with cold water to cool down the core.

Nuclear power accounts for eighteen percent of the electricity produced in the United States. Nuclear power is environmentally safe, that is, it doesn't produce acid rain or gases detrimental to the ozone layer. Nuclear power reduces our dependence on foreign oil. This will be more important in the future as our oil reserves become depleted. Despite the negative publicity, nuclear energy produces low radiation exposure to the general public.

On the downside, nuclear power produces highly radioactive waste which must be put in steel barrels and buried in caves or underground. This will remain radioactive for thousands of years and will be a burden to our grandchildren and their grandchildren will have to deal with it. As yet, there is no permanent waste disposal site in operation and the waste is being stored temporarily at many of the power plants.

Nuclear mishaps at Three Mile Island in Pennsylvania and Chomobyl in the Ukraine (Chernobyl in Russian) have made the public uneasy. At Three Mile Island on March 28, 1979 the nuclear core started to melt down and was exposed to air. A radioactive hydrogen bubble developed which was finally brought under control after a week of experimentation.

The Chernobyl accident occurred on April 26, 1987. In the accident the top blew out of the containment building and sent radioactive contamination to most countries of eastern Europe. This made much of the crops and livestock in the contaminated area unusable. However, many of the rural people local to the explosion were experiencing poverty and they consumed the affected crops.

There is no doubt that radiation increases the incidence of cancer, especially leukemia, bone cancer, lung cancer and skin cancer. Even low levels of exposure can result in an increase in birth defects. Further study needs to document the effects of extremely low levels of radiation over a long period of time, perhaps twenty to forty years.

Nuclear power plants are expensive to build and their waste products are expensive to handle and store. The idea behind nuclear power is enticing. A load of enriched uranium the size of a large desk will produce the same amount of electricity as three and half million tons of coal and there are no ashes to handle.

Leading Countries Operating Nuclear Energy Units in 2010
Nuclear reactors are used for military, commercial and medical purposes. The United States had 81 commercial units operating in 2010, 18 decommissioned units and 8 cocoon units which are created units not in operation. For instance, a unit in Connecticut met with so much opposition it never went on line.

There were about 450 operating commercial units in the world in 2010. Seventy two countries had nuclear reactors producing electricity. The leading countries in commercial nuclear power at that time were United States, Russia, Japan, France, United Kingdom, Germany and Sweden. Significant nuclear power was also produced in South Korea, Ukraine, Pakistan, Belgium and Canada.

Schemes for Alternate Energy Sources
Ancient people worshiped the sun and why not, when it comes down to it, all energy and all life depends on the sun. The sun is responsible for the hydrologic cycle which gives us running streams, the wind, the tides and plant growth. It is obvious that we are too dependent on fossil fuels for our supplies of energy. The following is a brief discussion of alternate possibilities for energy which will reduce our dependency on fossil fuels.

Solar Energy: We can heat our house by having large windows and letting the rays come through. This is a simple passive solar system of energy. We can also let the sun shine on water heaters and circulate the water to bathe in. The sun will heat stones during the day and they will radiate heat at night to keep us warm. There are many adaptations of passive solar energy and one has only to think about them a while to realize there is much potential there.

Active solar systems utilize solar hooked up to solar cells. The most common solar cells used today are found in hand calculators. Industrialized countries of the world are busy trying to perfect the solar cell that will provide electricity cheaply. Leaders in this research are the scientists of Japan., Italy and Germany.

Solar electric cells could provide much of our electricity of the future. Perhaps these can be used in the high sun months of summer and the conventional power systems used in winter thus conserving our fossil fuels.

Solar cells are easy to install, they take little maintenance and once installed will last for at least thirty years. Although solar cells are mostly silicon, which is an abundant mineral, they contain small amounts of cadmium or gallium which are rare. At present the cost of solar electricity is almost twice that of coal produced electricity. In the future the cost should become more competitive.

In 2011 Japanese research scientists claim to have detected large deposits of rare metals in the deep ocean which could be used in solar cells. We certainly hope they are correct.

Solar powered cars might be the wave of the future. Each year, engineers develop new solar powered cars and race them somewhere in the world. The winner of a race in Australia moved his car more than two thousand miles at an average speed off forty miles per hour.

Wind Power: Windmills have been used for centuries to grind grain and pump water. Today the wind can be used to generate electricity. Investigation into wind power became more intense in the 1970's and by 1987 there were almost 14,000 wind generators in the state of California which produced enough electricity for a million people. California hopes to provide at least ten percent of its electricity by wind power.

Water of Sea and Stream: There are many experimental electric generators hooked up to various coasts with exceptionally high tides. Basically as the tide comes in, it turns a generator as it pours through a narrow inlet. As it exits the shore the generator is reversed and the outflow again turns it in the desired direction.

Tide electricity facilities are expensive to build. And, it is possible that they will be subjected to severe storms. The few generators that exist have been very successful and this holds some promise for the future.

Experiments with wave generation of electricity have proved that it can be done. Several generating stations in France which were expensive installations are presently generating electricity. This form of power will have to be considered experimental and presently not feasible.

Hydropower, that is, the movement of water in streams has been used to grind grain, saw timber, power textile mills and move many types of other machines. Today we have many great hydropower installations producing electricity around the world.

Hydropower is inexpensive and it is pollution free. Once built the installation takes few workers to run it. The problems though are many. They are expensive to build and those in operation are presently having their water reservoirs filled with sediment. In the construction of these, good farmland and wildlife habitat are usually buried.

The Earth's Storehouse of Heat: With the invention of the heat pump it has been possible to extract heat from the air and water. The heat pump contains a gas such as ammonia which condenses and evaporates at reasonable temperatures. Even what we might consider to be cool air at thirty five degrees Fahrenheit has much heat in it. This air is brought into the heat pump, heat is extracted from it and the air is sent back to the outside.

The same is true for ground water. Water in the earth is heated by the earth's heat. This water can be pumped into a heat extractor pump and the heat removed and the water then sent back down into the earth to be reheated.

Heat can also be removed from ocean water under the same conditions. It is possible that we can have heat extractors riding on the ocean and have the heat or electricity created by the heat sent to shore.

The problem with heat pumps is that they take electricity to operate. If they are used to create electricity then there is no gain in energy. It will take as much energy to run the system as it creates.

Geothermal energy has proven to be more rewarding. Enormous amounts of heat are stored deep underground. This can be the result of small amounts of radioactivity or where geological activity is taking place such as along fault zones. These hot rock zones are capable of boiling water and when this water comes to the surface it shoots out forming geysers or runs out creating hot springs and streams.

There is a lot of subsurface heat in volcanic areas. A few countries such as Japan and Italy have tapped into this heat source.

Besides the obvious hot zones such as volcanism and earthquake areas, geologists have identified geothermal convection zones where the natural heat of the earth migrates in a predictable convection pattern.

The capital city of Iceland, Reykjavik, has used earth's thermal heat for years. The steam from an active volcano is used to generate electricity and the steam is used directly to heat houses in the city. Steam heat is also used in Iceland to heat greenhouses which supplies the city with out-of-season vegetables.

Volcanic steam is also used to generate electricity in New Zealand, Russia, the Philippines, Italy, Mexico, Japan and the United States.

In the United States geothermal electricity is produced in California, Montana, Wyoming and North Dakota. California has the biggest potential for producing large amounts of geothermal electricity. The generating plants are cheaper to build than conventional coal-fired or nuclear systems since there is no need for a source of fuel. By the year 2030, California could be generating one fourth of its electricity with geothermal energy.

Wood and Biomass Potential: Half of the world's population depends upon wood for its main source of cooking and heating fuel. Most of this takes place in the poorer Third World Countries, a.k.a. Lesser Developed Countries (LDC). But even in countries such as the United States, Sweden, Norway, Finland and Canada more than ten percent of home heating is carried on with wood.

Unfortunately in LDC countries wood resources are being depleted faster than they are grown. Most trees are gone from the fringe areas of the deserts and other low rainfall regions. In areas such as the plateaus of Latin America, Himalayas and on many islands such as Haiti the wood resources are almost gone.

The depletion of wood resources creates enormous problems for the environment. These include desertification, erosion, flooding, and habitat destruction.

The LDC, and indeed every other country must be taught the philosophy of sustained yield and conservation. Efficient cooking and heating stoves have been designed and these should be made available to villagers who now cook over open fires.

Everyone must learn to substitute cheap energy sources for those more expensive. Wind power can be used instead of electricity in many areas of the world to pump water and grind grain.

Biomass refers to any vegetable or animal matter. Biomass has not really been considered as an alternate source of energy for the world. We can create methane gas from decaying vegetation and manure. The gas can be used to heat buildings and run electric generators. This is being done in rural India where there is an abundance of manure. However, the manure used in this manner is taken from the fields that desperately need the fertilizer.

Ethanol, a gasoline type product, is being made from the distillation of grain and sugar cane. Ethanol can be mixed with gasoline to create a product called gasohol which is effective in running conventional motor vehicles. Brazil which has an abundance of sugar cane runs about thirty percent of its vehicles on ethanol. In the next twenty years it expects to run all of its vehicles on pure alcohol made from sugar cane. However, there have recently been big cutbacks in this program since small deposits of oil has been discovered in its territory.

Unfortunately, the creation of ethanol takes as much energy to produce as it creates. We are merely converting one form of energy to another and there is no gain in the transfer. If solar or wind power is used to operate the distillation apparatus then it is possible to gain energy in the creation of alcohol and ethanol.

Conserving Our Energy Resources

During the Energy Crunch of the 1960's and 1970's the United States gave much attention to the problem of conserving our energy resources and finding alternative energy. Once gasoline became inexpensive again the country forgot about its long range plans and returned to a live-for-the-moment philosophy.

Since the automobile takes most of our energy resources it seems logical to begin there. We can enforce car pooling and riding on public transportation as well as develop vehicles that double the present miles per gallon of gasoline.

Houses can be made energy efficient by insulation and using more energy efficient appliances and heating systems. Simple changes such as more efficient light bulbs and shower heads can make a tremendous difference in the total energy consumption if everyone used them.

We can change our philosophy as to what are necessities and what are luxuries. For instance, the United States uses more energy to run its air conditioners than the entire energy use of the country of China. The extra lighting in the city of Las Vegas, Nevada uses more electricity than a conventional city of five million people.

By lowering the room thermostat from 72 to 68 degrees there will be a fifteen percent decrease on the electric bill throughout the course of a winter. Savings can also be made by checking the house for drafts and filling these areas with caulking or weather stripping. Turning down the setting of water heaters to 120 degrees is also a cost cutting move.

Electric Cars

Cars cause more air pollution than any other source. The two most promising energy saving sources for cars of the future are electric from batteries and the use of hydrogen gas. Some states are moving to a standard of auto emissions that will make driving with gasoline an expensive enterprise.

Electric cars give off no emissions. store electricity in batteries and use the batteries to run electric motors. Electric cars do not contribute to urban smog or greenhouse warming, at least not directly.

At the present time, most electricity in United States is produced by burning coal. Since these coal power plants are in rural areas the net effect will be to reduce urban smog.

However, the power plants which burn fossil fuels are twice as efficient as vehicles which burn gasoline. So for each thousand miles traveled the net effect is less pollution.

Batteries are not yet available that will make electric cars feasible. The experimental batteries have a high cost of production and use toxic materials in their construction.

14. HARMFUL SUBSTANCES

From 1980 to 1995 more than 400,000 deaths a year in the United States were attributed to the greatest environmental hazard to human life. What was the hazard? It was the use of tobacco. In that same time period, automobile accidents killed about 45,000 a year and hard drugs 30,000. A twenty year old man or woman in good health who uses tobacco cannot imagine that thirty years later they might be dying from some activity that did not seem to be harmful at the time.

A hazard is any substance that can cause disease, injury or environmental damage. Hazards are measured by statistical analysis and computer models tell us the probability of the hazard affecting our lives.

The effects of the hazard depends upon the type of hazard, perhaps a chemical in the air, how long the time of exposure, the health and age of the individual exposed and how well parts of the body respond to eliminating the hazard. If the effects are immediate it is labeled acute and if it is a long lasting effect due to exposure over a long period of time it is a chronic effect.

Cultural hazards are those produced by our living and working conditions and by our activities such as drinking alcoholic beverages. This would also include our attitudes toward driving and unsafe sexual activities. Pollution is usually the result of cultural activities.

Physical hazards include weather phenomena such as tornadoes and earthquakes. These would also include natural fires and noise. Mountain climbing is a cultural activity that includes physical hazards

Hazardous Environmental Substances

There are thousands of chemicals and other substances in the environment that can be dangerous to human health. We need not worry about most of them since they exist in amounts too small to be a threat to humans. However, some of them exist in large quantities and in these quantities can be hazardous. Some others, even in small amounts can be deadly. A few of these are listed below.

Carbon Monoxide

When we breathe in carbon monoxide, it combines readily with our blood. It combines more than 200 times faster with hemoglobin than does oxygen. Breathing in even a small amount of carbon monoxide for a full workday decreases oxygen in the body equivalent to a loss of one pint of blood.

Carbon monoxide is colorless and odorless. When we are in a lot of traffic, carbon monoxide build-up can be a serious threat to our ability to reason.

People with heart disease, asthma and lung disorders are especially vulnerable to carbon monoxide poisoning. Carbon monoxide coupled with high altitude and high humidity are double dangerous.

Cigarette smoke contains 300 parts per million of carbon dioxide. This reduces hemoglobin oxygen in the non-smoker as well as the smoker. Cigarette smoking along with heavy traffic and industrial pollution is a serious threat to life.

Lead

Lead poisoning causes mental retardation in children. Adults have the ability to pass off excess lead in their urine. However, in adults, lead coupled with high blood pressure can be dangerous.

High levels of lead in humans were attributed to leaded gasoline and the fumes resulting from its combustion. It was also found that the gaseous lead settled on plants and that such things as garden vegetables grown in high traffic areas had high amounts of lead on them. This lead on fruits and vegetables is easily removed by washing. Prohibiting lead in gasoline has corrected the situation in most instances.

Peeling paint in rural areas was responsible for several documented cases of lead in groundwater supplies. Peeling paint in older houses is believed to be the source of excess lead in school children. Most school districts now check for lead in children. Lead found in drinking water is now the most serious threat from this substance. This comes from lead solder found in most water pipe connections.

Airborne lead can be the result of burning slick magazines. The paints and dyes used in the magazines have other heavy metals besides lead.

The Environmental Protection Agency has banned the use of leaded gasoline but it is still manufactured and used in such things as chain saws and lawn mowers. We should suspect any anti-knock additive to gasoline. These substitutes for lead may be just as damaging to the environment and human health as lead. Any gasoline gives off a plethora of undesirable gases when burned.

Asbestos

Asbestos is a fibrous mineral heavily mined in Canada, Russia and South Africa. It is known as mineral wool and has been used in fire prevention and heat retaining structures. Major uses are in furnaces, face masks, hot pipes, oil lamp wicks, ironing board pads, brake linings, plasterboard and plaster.

As asbestos ages, small flakes get into the air and into the lungs. These act as tiny knives that cut the lungs. The lungs respond by growing scar tissue around the cuts. These appear as red-yellow growths in the lungs and these reduce lung breathing capacity. The disease is called asbestosis.

Asbestos lungs are further aggravated by cigarette smoking. Lung cancer deaths are extremely high among asbestos workers that smoke.

There is a national campaign to remove asbestos from all schools. Older schools have asbestos in flooring material, plaster and ceiling tile and covering for water and furnace pipes. However, research indicates that there have been no harmful effects from asbestos in these school construction items as long as it remains undisturbed. Perhaps the billions of dollars spent on removing asbestos was an unnecessary expense.

Mercury

Mercury is a metal found in liquid form at normal temperatures. Its deadly effects were recognized in ancient times and it was a common substance used in murders and suicide.

About five thousand tons of mercury are put into the environment by mining and industrial uses of the mercury ore, cinnabar (HgS) and the burning of coal.

Inorganic mercury and its vapors do no pass through cell tissue easily. It damages liver, intestines and kidneys. Organic mercury (methyl mercury) passes easily through cell tissue and is readily absorbed in the food chain. When humans eat fish contaminated with mercury, then mercury poisoning results. It eventually results in irreversible nerve and brain damage. If the intake of mercury ceases before a critical point is reached then recovery will result.

In the Japanese town of Minamata, severe mercury poisoning resulted when its citizens ate contaminated tuna and swordfish. The poisoning was first recognized when birds fell from the air and were eaten by cats. The affliction was called the "disease of the dancing cats".

The illness now known as Minamata Disease first attacked the families of fishermen. Its first symptoms were fatigue, irritability, headaches, numbness in body parts and subtle hearing loss. Eventually vision became blurred. Babies were born with congenital defects. Before the disease was officially identified 43 people had died and were severely disabled.

Other heavy metals which are hazardous to human health include arsenic, bismuth, cadmium, chromium, copper, gold, lead, nickel, platinum, selenium, silver, thallium vanadium. Each of these has important uses in our modem society. They escape into the environment during mining, manufacturing and final disposal. Less than twenty years ago a common method of disposal of heavy metal wastes was to dump them into lakes, streams and the ocean. The metals are still there today and clean-up of these is almost impossible.

Dioxin

There are about 80 known dioxin compounds that are distinguished from each other by the chemical arrangement of the chlorine atoms in a molecule. Dioxins contain hydrogen, oxygen, carbon and chlorine. The most deadly of the dioxin molecules is TCDD.

Dioxin is a by-product of chemical reactions in the manufacturing process and also the result of smokestack burning at low temperatures. It is the dioxin in pesticides, especially Agent Orange, which is the lethal ingredient.

Exposure to dioxin produces a condition known as chloracne (chlorine-acne) which may be accompanied by loss of appetite and weight as well as liver disorders and nerve damage. Exposure to even small amounts of TCDD will result in eventual cancer.

Once formed, dioxin persists in the environment and there is no feasible way of getting rid of it. In 1983, the town of Times Beach, Missouri was contaminated with dioxin from oil sprayed on its gravel roads to alleviate a dust problem. The federal government purchased the entire town and evacuated its 2400 residents. There are probably more than a thousand sites in the United States contaminated with dioxin as well as many bodies of water, such as the Great Lakes.

Pesticides

Ever since humans arrived on earth they have been bothered by pests in various forms. These have given humans competition for food, threatened human health and some have simply made life miserable.

To control the pests, various ways of killing or abating their numbers have been devised. The first pesticides were various forms of sulfur, lead, arsenic and mercury. When the New World was invaded by Europeans, they adopted nicotine and other plant derivations such as curare into their pesticide arsenal.

The recent expansion of modem pesticides can be traced to World War II when DDT (Dichlorodiphenytrichloroethane) was invented. It killed disease spreading and plant eating insects by the billions. Cows were sprayed with it to eliminate ticks and since it persisted on the animal it killed flies that landed on them.

About twenty years after the invention of DDT, it was discovered that it caused a considerable amount of environmental damage. DDT collected in fatty tissue of humans as well as animals. Birth defects were connected to DDT. If a human ate a fish containing DDT, the chemical became part of the human. DDT was especially disastrous to wild birds, such as the Bald Eagle, which feed on fish which feed on water organisms. Eventually DDT was banned from use in the United States. However, it is still used in many Lesser Developed Countries.

Today, there are about 60,000 different chemical pesticides on the market. These are sold mostly to kill insects, weeds, rodents, fungi, mites and ticks. We associate pesticides with farming but homeowners used about five times more pesticides per acre than farmers. One fifth of our pesticides is used on lawns, golf course, parks, gardens and cemeteries. More than a billion and a half dollars a year is spent on lawn care in the United States.

Pesticides include insecticides rodenticides, herbicides and fungicides. Each has its own target specialty. Each has its advantages and disadvantages.

Insecticides are mostly phosphate, carbonate and chlorinated hydrocarbon compounds which includes pyrethroids, heptachlor, toxaphene, chlordane, kepone and mirex. Many of these stay in the environment for at least twelve years. They function by damaging the insect nervous system. These insecticide chemicals are slowly being replaced by others which have a shorter persistence in the environment.

Phosphate insecticides have a persistence measured in weeks but they are particularly harmful to humans, birds and fish. Phosphates are water soluble so they are more likely to contaminate water supplies. Since phosphate insecticides only last a few weeks, farmers are more likely to use them several times over the period of one growing season. Malathion and Parathion are the most used phosphate insecticides.

Carbonate and pyrehroid pesticides are active only a few days, but extremely harmful if improperly applied.

Rodenticides are used to kill rats and mice. Most of these are sodium fluoroacetate which attack the nervous system. Other rodenticides, sold over the counter and applied in houses, cause rodents to get thirsty and go outside in search of water. Once outside, they hemorrhage and die. \

Herbicides are active for only a short time. In that time they are used as defoliants or to sterilize the soil. Contact herbicides kill the plant quickly. Others are taken into the plant vascular system to eventually kill it. The popular contact herbicides are triazines, atrazines and paraquat. The system absorbing chemicals are 2,4-D and 2,4,5-T. The latter, which contains an ingredient called Agent Orange, has been banned in the United States.

Herbicides which sterilize the soil kill micro-organisms essential to plant growth. Most of these also get into the vascular system of plants which kill them and eventually make the soil barren. Later this sterile area can be planted and fertilized to produce crops.

Potato plants are sprayed with a vegetation killing substance before the harvest. The dead plants dry out quickly and then the potatoes are scooped up by toothed unearthing machinery without being clogged by vegetation.

Fungicides are mostly used to protect seeds before planting. These use arsenic, lead and mercury. Humans eating the contaminated seed can develop a number of afflictions and death is common. Humans have become ill and died from eating hogs and chickens which had eaten contaminated seeds. Most treated seeds are now dyed purple or red in order to warn the user that a fungicide has been applied.

Fungicides are also used to counteract human fungi afflictions such as ringworm, athlete's foot and mildew. Fungus skin infections are extremely painful.

The Case for Pesticides

Pesticides are mobile. They can be applied in one area of the world and end up thousands of miles away. They are carried by air and water. They are carried in the bodies of fish and birds. They stay in water sediments indefinitely. When the sediments are dredged, the pesticides enter the environment once again.

Pesticides destroy beneficial insects as well as harmful insects. They cause resistant strains of insects to develop. If two percent of sprayed flies survive, they will unite to foster a new super breed of flies.

The biological magnification of pesticides is one of society's horrors. A fish eats the contaminated larva which ate the contaminated plant. We eat the fish. It is logical then to eat only small fish instead of lunkers, since they have less contaminants. Fish stocked by hatcheries are usually free of contaminants.

However, before we condemn pesticides we must keep in mind that insects kill over a million people a year and cause another 300 million people to become ill. Mosquitoes spread malaria, yellow fever and encephalitis. Ticks spread Lyme Disease and spotted fever. Fleas spread bubonic plague and typhus. Various flies spread hundreds of different diseases to both humans and animals.

Consider also that insects eat about thirty percent of the United States crop and fifty percent of the world crop each year. This is before and after harvest. When we consider this insect devastation coupled with loss to weather, birds, weeds and mammals, it is a wonder we can produce the amount of crops we do.

The ideal pesticide would be very selective and kill only its target. It should have a short life in the environment and decompose harmlessly. It should save more money than that lost to the pest and it should limit the development of genetically resistant pests. So far the ideal pesticide does not exist.

Farmers can limit the use of pesticides with crop rotation, strip cropping and developing trap crops which is a crop that takes the insect away from the desired crop.

For best environmental results we should consider Integrated Pest Management (IPM). This is a method of reducing pests with natural factors that help manage them. The goal of IPM is to reduce the size of the pest population below the injury level rather than to eliminate the pest entirely. Perhaps the best control is picking the pest off the crop by hand and simply stepping on it.

Consider the following methods of pest control. Use proper plant selection. Choose varieties of plants that are well adapted to the climate. Some are more resistant to diseases than others.

Handpick, shake or wash off pests rather than spray them with pesticides. Use barriers such as screens or insect repellents to ward off insects. Use traps for flies, mice and beetles.

Many newer insect traps contain insect sex attractants. Where possible we use the sterile male technique, that is releasing sterilized male of the species to mate with receptive females, thus there are no offspring.

Put up bird houses to attract flycatchers, bluebirds and swallows. Use BT, Bacillus thuringeinsis to kill larva. This is a biodegradable bacteria which is harmless to humans. It is used to kill caterpillars on cabbage type plants.

Is Your Lawn Worth The Risk?

Lawn care chemicals are designed to fertilize lawns and kill unwanted plants, insects and fungus. The poisons used in these mixtures include herbicides such as 2,4-D, insecticides such as Dursban and fungicides such as Captan.

Long term problems associated with lawn care poisons include lowered male fertility, miscarriage, birth defects, chemical sensitivity, liver and kidney damage, heart disturbances and cancer. Immediate adverse symptoms include depression, anxiety, irritability, vomiting, dizziness, fatigue, headache, diarrhea, coughing and asthma-like attacks.

There are no laws to protect us from being poisoned by lawn care chemicals. According to congressional records nine out of ten pesticides in use today were registered with health testing that is non-existent, incomplete or fraudulent. It is a violation of federal law to label any chemical pesticide as "safe", "harmless", or "non-toxic to humans or pets".

Most doctors are not familiar with the symptoms of chronic low-dose pesticide poisoning. This condition is most often diagnosed as allergies, asthma or the flu.

Lawns can be healthy if they are planted in a natural environment, not in a desert or some area where grasses do not exist. Leaving grass clippings on the lawn is the easiest way to handle them and it adds fertility to the soil. Leaving clippings on the lawn helps build a community of microorganisms and earthworms which help to keep the lawn healthy. Many lawn mower companies now advertise the "mulching" lawn mower which shreds the grass to small size and leaves it on the lawn.

Major Crops Treated With Pesticides

Insecticides: 1. pears 2. apples 3. citrus 4. almonds 5. grapes
Herbicides: 1. rice 2. peanuts 3. corn 4. cotton 5. Sorghum
 6. soybeans

Common Chemical Contaminants Dangerous to Drinking Water

arsenic - nervous system
benzene - genetic material
cadmium - kidneys, bladder
carbon tetrachloride - liver, kidneys, lungs
chloroform - liver, kidneys
dioxin - causes many cancers
ethylene dibromide - male sterility
lead - birth defects
mercury - nervous system
nitrates - respiratory system
PCBs - liver, kidneys, heart
trichloroethylene - nervous system, liver, kidneys, skin problems
vinyl chloride - liver, kidney, heart, stomach

15. WASTE DISPOSAL

Everything you are wearing and everything in the room with you will eventually become waste. It will all have to be thrown away or disposed in some manner.

The United States contains less than five percent of the world's population but produces one third of the world's solid waste. Our per capita solid waste amounts to 44 tons a year. This comes, not only from individual daily throwaway waste, which amounts to four and a half pounds per person, but from industry and agriculture which manufacture products we use.

About eighty five percent of the United States waste is generated by mining and industry. By tonnage, most of this is mine waste which is piled around the mine site.

The second largest creator of solid waste is agriculture which produces large amounts of stems, vines, stalks, twigs, leaves and manure. Industrial solid waste is third in landfill weight. This includes scrap metal, plastics, paper and sludge. A large amount of fly ash is generated by coal fueled electric plants and municipal incinerators.

Municipalities generate 185 million tons of solid waste a year. Each individual person generates about 1600 pounds of garbage a year. Most of this ends up in landfills. Many municipalities such as Philadelphia pay over a hundred dollars a ton to use present landfills.

Broken down by producers the solid waste generation for the United States for 2011 was Mining 75%, Agriculture 13%, Industry 10% and Municipal 2%. Most of the municipal waste ends up in landfills, 73%, but incineration is increasing, 14%. Another 12% is recycled and only one percent is composted. Efforts are being made to increase recycling and composting.

Paper and cardboard make up about a third of household waste. This includes newspapers, paper towels, napkins and food packaging. Much of this could be recycled but adequate paper recycling plants are still a few years away.

Municipal solid waste in the United States in 2011 included paper 39%, yard wastes 17%, food 9 %, glass 8%, wood 3%, leather and cloth one percent. Almost all of this can be recycled or composted.

We usually deal with the growing waste pile by burning or burying it but it would be better if we cut down on production of materials which will become solid waste in our throw-away society. Most packaging of food, tools, toys and the like products come with too much wrapping. Even though this has benefits in handling and retailing it is waste of valuable resources.

Burning solid waste, such as paper, causes air pollution. Even the best designed incineration systems produce harmful airborne substances. Burning plastics may create dioxins. The waste ash of incineration is also toxic and must be disposed with care, usually in a hazardous waste dump. These waste dumps are always a threat to precious water supplies.

To help decrease the amount of solid waste generation we can give tax write-offs to industries which consider the environment in their operations. Our present system of tax incentives to industry encourages production at the expense of the environment.

We must encourage low waste production and emphasize recycling and reusing waste products. Around eighty five percent of materials put in solid waste landfills can be reused or recycled. Eventually, however, everything, even the recycled products, ends up in a landfill.

Three fourths of our municipal solid waste ends up in sanitary landfills. These are garbage dumps where solid wastes are spread out in layers, compacted with bulldozers and then covered with plastic or clay each day. Modern landfill regulations require the final filled landfill to be covered with clay or some impervious material. Many of the landfills are also lined with clay or plastic to prevent leaching into ground water supplies.

Since modem landfills are covered and compacted, the normal breakdown of materials by biologic processes is severely curtailed. Newspapers in a modern landfill are still readable after thirty years. Plastics may last up to 200 years before they break into small pieces.

Landfills with a lot of organic materials will produce methane gas. These landfills can explode or catch fire and smolder for years. In many landfills, gas venting pipes are inserted into the landfill to take off excess gas. The gas can then be collected and used to run machinery.

Solid waste not in landfills is usually incinerated. This has several advantages over land filling. It kills disease organisms and reduces the need for landfill space by ninety percent. However, it does not discourage the throw-away habit which causes many of the landfill problems in the first place.

Incineration may reduce landfill space use but it pollutes the atmosphere. Also, the residue of incineration contains concentrated toxics which must be put in landfills with care.

Many modern incinerators produce electricity while burning the trash. These are usually located near cities to take advantage of the large trash generating population. This is also where air pollution from burning does the greatest damage.

Also on the negative side of incineration, is the fact that recycling paper will save five times more energy than is generated by burning. Incinerators are very expensive to build. Even the most efficient of them produce small amounts of dioxins, acids and airborne heavy metals such as mercury, lead and cadmium.

Most conservationists believe we should limit solid waste at its source rather than deal with it as the end product. But, at the end product we can separate trash into recyclable and non-recyclable materials. The recyclables can generate income for those involved in its collection.

Much of what is thrown away can be reused by someone else. It is a matter of finding that ultimate consumer. Goodwill Industries and the Salvation Army do a good job of collecting used furniture and clothing for redistribution. What used clothing collectors cannot immediately redistribute in America is usually sold by the pound to Third World importers. This greatly reduces pressure on our landfills.

Biodegradable yard wastes should be composted and used for soil improvement and fertilizer. This can also be done with slaughterhouse and food processing wastes as well as feed-lot manure. Compost can be produced at large recycling centers, bagged and sold at a profit. Households should have their own compost bins and the compost products used on flower beds, lawns and gardens.

Today, landfill space is becoming critical and recycling, incineration and composting will increase as the need to reduce landfill space gets even more critical. In order to reduce the volume of material going into landfills we need beverage container laws requiring the return of containers for recycling or refilling. Old tires, a quarter billion a year in the U.S., can be melted into crude oil or ground up for parking lot surfacing.

A tax could be placed on throw-away products and the money used to take care of that product's final disposal. For instance, every new automobile would come with a hundred dollar tax. The money would be put into a fund and used to dispose of the old automobile carcass when it has expired. A ten cent tax on all tin can products would accomplish the same purpose. There has been an increase in people who will buy junked automobiles since they have many parts that are made of metals with value. However, there are still parts of the automobile that need some special care in disposition.

We can move a long way in reducing waste by reducing packaging. One dollar in every ten spent in grocery stores is for packaging. Many products such as toys need no packaging.

Manufacturers should be encouraged to make products which are easy to recycle and reuse. Consumers can encourage this by changing their purchasing habits.

Everything we own or use will eventually end up as a waste product. With environmental sensitivity we can limit our contribution to the waste stream. We can't put off the inevitable landfill shortage but we can slow it down for the present.

Hazardous Waste

Much of our waste products are considered hazardous. These are materials that contain substances which are toxic, carcinogenic and genetic threats. Hazardous also includes flammable and explodable materials as well. It also includes materials that break down to form dangerous substances such as heavy metals.

Each hazardous material has its own disposal problem. Usually the Environmental Protection Agency classifies the wastes into categories according to the most appropriate method of treatment and disposal.

Aqueous Liquid Wastes include those that are corrosive such as pickling baths and cyanides. **Organic Liquid Wastes** include halogenated solvents, oils, combustible liquids and other solvents. **Solids and Sludge** include paint residues, combustible solids and sludge, heavy metal sludge, electric arc furnace waste and emission control dusts. The category, **Other Wastes,** include ignitable, corrosive, reactive and toxic materials.

These are industrial wastes and there are many methods devised to handle them. However, much of this material is put into landfills rather than recycled, neutralized or reused.

Hazardous waste recovery facilities use solvent recovery techniques. These separate the contaminants from solvents and thus restores solvents to their original quality or lower quality for reuse.

Pyrometallurgical Recovery uses high temperatures to separate metals from materials, mostly ores. Hydrometallurgical Recovery concentrates heavy materials such as chromium and nickel by various weight separation processes. Acid Regeneration involves separation of unused acid from pickling baths. Fuel Blending creates usable products by taking used oils and solvents and blending them with fresh materials having a high BTU value.

Hazardous wastes can also be treated to neutralize their hazardous properties. Incineration is a major treatment. Wet Air Oxidation involves oxidation reactions identical to combustion but occurring in a liquid state with elevated pressure and temperatures. Advanced Thermal Destruction which uses elevated temperatures as the primary means of treatment has not yet been perfected.

Land disposal systems involve surface impoundments which holds the materials until other means of disposal can be implemented. There are about 78,000 of these in the United States and there seems to be no great rush to treat the materials.

Landfill disposal involves permanent emplacement of hazardous waste. Nobody seems to want this type of landfill in their neighborhoods and we have developed the NIMBY CONCEPT which stands for "Not in my backyard". We use the products that generates hazardous wastes but we don't seem to want to be part of the disposal of the wastes created by those products.

There are many federal regulations covering hazardous waste landfills. These involve prohibition of certain wastes such as those with high radioactivity, use of plastic or clay liners, leak detection, leachate collection, impervious caps, ground water monitoring and perpetual maintenance.

Eventually the plastic and clay liners will begin to leak. By that time, the owners of the landfills will be out of business and there will be no responsible parties to clean up the threat and the cost of clean-up will fall back to the federal government.

Let us not get the impression that only industry puts hazardous materials into landfills. We have only to look in the mirror to see someone who uses hazardous materials and must dispose of them. For instance our hazardous materials which go into landfills include batteries, antifreeze, motor oil, rust removers, disinfectants, toilet cleaners, oven cleaners, spot removers, oil base paints, paint strippers, wood preservatives, driveway and roof tars, pesticides, flea powder, glues and deodorants. These are not necessarily hazardous unless they get into our water supplies. Once in a landfill it is only a matter of time until they get into the water system.

Low Level Radioactive Waste Disposal

Putting low-level radioactive and hazardous waste into landfills instead of the secure radioactive waste disposal is a cause for much debate between environmentalists and those who produce wastes. Low-level radioactive waste are such things as rags, papers, protective clothing, filters, gloves which comes from making electrical energy by nuclear energy, from medical facilities and industrial and medical research. It is also created by the manufacture of smoke alarms and other warning devices.

Low level wastes do not include uranium rods from nuclear power plants and high radioactive waste from weapons manufacturing. These are handled by the federal government at separate waste disposal sites.

Low-level waste is classified according to the amount of radioactivity it contains. States are responsible for the disposal of low radioactive wastes. Disposal sites will not accept low-level wastes in the form of liquids or gases, only solids.

Each state must take steps to provide for the safe disposal of Low-Level wastes generated within their borders. If a state does not provide for the disposal, the state may be barred from using any of the present disposal sites operated under federal jurisdiction.

High level radioactive disposal involves plutonium which is an element that exists in trace amounts in nature but is manufactured in large amounts in nuclear reactors. Plutonium is created by bombarding atoms of uranium U 238 until they absorb a neutron and become Pu 239. The Department of Energy operates thirteen sites for storage and disposal of plutonium. These storage sites which contain about 28 tons of plutonium are aging and possibly pose a threat to their surroundings.

The plutonium storage sites are located at Hanford WA, Argonne West ID, Rocky Flats CO, Pantex TX, Los Alamos NM, Sanda NM, Lawrence Livermore CA, Lawrence Berkeley CA, New Brunswick IL, Argonne East IL, Mound OR, Oak Ridge TN and Savannah River SC. All of these are classified as temporary disposal sites, but we do not yet have an operational permanent disposal site for high-level radioactive wastes.

Superfund

A fund established by federal and state governments was created to clean up inactive hazardous waste dump sites around the

United States. It was first initiated in the Comprehensive Environmental Response Compensation and Liability Act of 1980. It has since been known as Superfund.

In the first phase of the act 34,000 hazardous waste sites were identified. This included almost 18,000 sites at military bases. Today, there have been almost half a million of these sites identified and the number keeps growing as these sites, usually hidden, are located.

Unfortunately, only about twelve percent of the original 16 billion dollars allocated for clean-up has gone for actual work. The rest has been used for administration, consultants and management. The entire clean-up costs will probably come close to a trillion dollars before it is completed, if it is ever completed.

Unfortunately again, many of those who have created the toxic waste sites are no longer in business and cannot be prosecuted. Also, many of them that can be prosecuted have political clout because of campaign contributions and other perks for congressmen. Clean-up has been slow and there are thousands of sites to go.

Meanwhile, people live on the fringes and around these sites. The Love Canal Site is a classic and well publicized example of the situation. Love Canal is located near Niagara Falls, New York. It was an old canal filled with hazardous wastes and covered over. A builder, William Love, sold the land to the Niagara Falls school district for one dollar. An elementary school was built on the site and residences were built near it. The area had a high incidence of birth defects, miscarriages, cancer, nerve and kidney disease. Eventually, the state closed the school and permanently relocated the 238 families who lived near the dump which they could not see.

The properties around the Superfund sites have become nightmares to the people who live near them. Most of them cannot sell or simply abandon their properties. They can only wait for the federal machinery to come to clean up the sites.

Not only do we have to eradicate these sites but we must take steps to insure that no new ones are created. Is your congress man or woman willing to cooperate in this effort? Since I have observed Congress for more than sixty years I would be surprised if they gave anything but lip service to this severe problem. Sorry to say, but congressmen are more interested in getting re-elected than in actually serving society.

16. FUTURE PROSPECTS

Present trends have some implications for the future. In the United States more members of the middle class are sinking into poverty. The population of the middle class is declining, poor people are increasing and more affluent (rich) people are getting more affluent.

Since 1980 (Reagan Years) the standard of living for the American family has been declining at the rate of four percent per year. There were more people living below the poverty line in 1995 than in the entire history of the country. There were more people unemployed in 1995 than during the Great Depression. In July 2011 the unemployed, those people collecting unemployment and registered for jobs was over ten percent of the work force. We are still the most envied country around the world but that status is rapidly eroding.

Most poor countries have high population increases. This results in migration to other regions. Even newly industrialized countries have population increases beyond the carrying capacity of their lands. Our neighbors are sneaking across our borders in larger numbers every year and putting stress on our social service agencies. In 2011 more than a thousand people a day made illegal crossing of our border from Mexico. During the Carter presidency more than a hundred thousand Haitians came to Florida in make-shift boats. At that time more than fifty thousand Cubans entered our country illegally. Boatloads of potential Chinese immigrants floundered and were rescued on both the Atlantic and Pacific coasts. They were given refugee status. This immigration, legal and illegal, coupled with our modest fertility rate will double the United States population in less than fifty years.

We have roughly 310 million people in the United States and by the year 2050 we could have over 500 million. This gives us a very short period of time in which to double our social services, our schools, our water supplies, our medical services, our food resources, our sewage systems and everything else necessary to our standard of living.

No one who considers the situation can deny that our quality of life is diminishing. Most thinking people will agree that our population has already reached, if not exceeded, the carrying capacity of the environment. This is true with most countries of the world.

What is our obligation to the rest of the world? Can we continue to accept immigrants as we have done in the past? Most countries have stopped immigration altogether and only the United States, Germany, Canada and Australia still accept other peoples in large numbers. Recently Germany has enacted legislation to halt the influx of foreigners, mostly refugees from war zones.

With no increase in productivity and a doubling of our population we will simply have to accept less out of life. We can look forward to energy shortages, exhausted land, scarce water and a radical change in our diets. Our lands are already at the limit of production and crop yield increases are not going to keep up with the population growth. We will no longer have the luxury of exporting food.

The basis for our farm productivity is petroleum. We will run out of domestic petroleum supplies in twenty years and foreign sources will most likely hang on to their limited supplies. We are losing farm acres to urbanization at the rate of about 2 million acres a year. Most "thinking" countries of the world have limited the development on arable land. The United States mentality seems to think that it is the individual's right in a free society to do what one wants with private property.

Today, the average American spends about fifteen percent of income on food. In Europe and Japan the figure is thirty percent and in Lesser Developed Countries the cost runs from fifty to a hundred percent of one's income.

The hope for our future depends on legislation and regulation enacted by the United States Congress. Eventually they will be forced to recognize the desperate needs of our future and we can only hope sensible legislation which limits the profit motive as a consideration is enacted.

We must limit our population growth and wasteful use of our resources. We must treasure our water, air and soil. We must conserve, practice sustained yield and recycle.

At the beginning of 2011 there were over fifty wars in progress. Most of these were not reported in the American press. We heard about Afghanistan, Chechen, Iraq, Tajikistan, Rwanda and Somalia. We heard very little about Mexico, Peru, Ecuador, Sri Lanka, Cambodia, the Philippines and Indonesia where intense fighting was carried on. The present drug war in Mexico has resulted in over a million deaths in the last four years.

Most of these wars were over resources and a major resource was water. In this country, skirmishes are shaping up over resources of land and water. The burgeoning population centers of Southern California have already exceeded the carrying capacity of the land. Groups of citizens living in Northern California have organized to break free of the south. How far these ideas will go largely depends on stress factors.

Los Angeles already transports most incoming water over three hundred miles. It uses about nine billion gallons of fresh water a day. Its population will double in about fifteen years. Where will it get the extra nine billion gallons of water EACH DAY?

In this country, skirmishes are shaping up over resources of land and water. The burgeoning population centers of Southern California have already exceeded the carrying capacity of the land. Groups of citizens, living in Northern California, have organized to break free of the south. How far these ideas will go largely depends on stress factors.

The international community has many organized meetings to discuss the implications of population, food, refugees, health and resources. Not much has been accomplished at these meetings and most countries favored their own agendas. However, what has been accomplished is the understanding that there are problems that have to be solved and most of these have to be solved on the international level.

THE EARTH SUMMIT - Brazil 1992

The most acclaimed international meeting the world has ever seen was held in Rio de Janeiro, Brazil in June of 1992. It was the United Nations Earth Summit on the environment which hosted delegates from 178 nations including 116 heads of state.

The world expected the United States to lead the way into an environmental awareness that would have far reaching effects well into the 21st century. Instead, the United States, under the leadership of President George Bush I, had a negative impact on the proceedings.

The United States insisted that carbon dioxide emission standards be eliminated from the wording of the global warming treaty. At the time of the treaty the United States with 5% of the world's people produced 23% of the world's industrial carbon dioxide.

The United States position was that the wording of the "greenhouse gas" emission treaty was too strict in its first draft and pushed for a useless high-sounding document which eventually let any nation oversee its own emissions. President Bush said his "obligation is to protect American jobs as well as the environment".

A biodiversity treaty was completely opposed by the United States and President Bush refused to sign the final draft. The treaty set aside important habitat areas in order to save and preserve diverse species. President Bush said that it would require additional U.S. aid to poor countries and would harm America's biotechnology industry.

Many countries will diminish their resources such as rainforests until their economies are improved by foreign aid. Without foreign aid they are forced to proceed with resource exploitation for their own economic survival.

Japan paved the way to the future when its Prime Minister Kiichi Miyazawa pledged $7.7 billion in environmental aid to developing nations. He stated that Japan would also eliminate the use of all CFCs in three years and reduce its carbon dioxide emissions to 1990 levels by the year 2000. Japan emerged as the environmental leader of the world. **Note:** Japan did do this.

It may be noted that at the time of the Summit, Japan fishing fleets still hunted whales, businesses still traded in endangered species and Japanese lumber imports were responsible for destruction of large tracts of tropical and temperate forests.

Basically, the main tenets of the Rio meeting were vague and the spirit of the meeting was its strong point. In the brief ten days, the world focused on the environment and perhaps in the future, all countries will be sensitive to the issues outlined at the meeting.

INTERNATIONAL CONFERENCE ON POPULATION AND DEVELOPMENT 1994

Cairo, Egypt was the meeting site for the International Conference on Population and Development sponsored by the United Nations. It was assumed by the sponsors that cooperation in limiting population and increasing opportunities for poor countries would be the agenda.

However, differences in culture became evident and opinions on how to improve the lot of most of the world differed.

Although the main topic was population it was really about **women's rights**. The original draft proposals called for a full range of reproductive and health-care services, including contraceptives and sex education for women. The organizers believed that the equality of women was the cornerstone of any population program.

The premise seemed simple enough, "what happens to a woman should be her decision". This might have been fine for most western cultures in highly developed nations but applied to other cultures it was an explosive issue.

Some Muslim clerics, from countries where women are hidden by veils and where the Koran's teachings are the law of the land, were outraged at the idea of equality for women. The agenda was an attack on their culture.

Giving the Islamic protests extra support was the Vatican led by Pope John Paul II. He accused the United Nations of trying to establish a worldwide right to abortion and demeaning the importance of the family. Catholic priests attacked the draft program for allegedly encouraging homosexuality and adolescent sex.

The American delegation was led by Vice President Albert Gore who was recognized around the world as a concerned environmentalist. He and his staff were convinced that population limitation is the world's central issue.

The American's pointed to Thailand as an example. Thirty years previous to the conference it was a poverty stricken country. Its government began to stress the importance of women's health and education. Female literacy increased to 90 percent. The average number of babies born to each woman decreased from six in 1968 to 2.1 in 1994.

Many interesting statistics were presented at the meeting. For instance, of the 960 million illiterate adults in the world, two thirds are women. Of the 130 million children denied elementary education 90 million are girls. Women are legally beaten almost everywhere.

Food per capita is falling around the world and the population continues to increase. Raising the status of women seems to be a way to counteract the dilemma. It is a tragedy that in many areas of the world women are treated as property rather than as an integral part of humanity.

On a personal note, I gag when I attend a wedding and the official in charge asks "Who gives this woman to this man?"

In most areas of the world there are more people than the environment can support. This leads to poverty which leads to destruction of the environment. Each region has a carrying capacity and when this capacity is exceeded, extreme measures must be implemented in order for the population to survive. Food and other resources must be imported or some of the population must move out. If this cannot be accomplished, the usual result is war.

Many environmentalists believe the population of the United States has reached its carrying capacity and we are already importing many resources to compensate for the excess population. For more and more Americans, the standard of living continues to diminish. Today, there are more Americans living below the poverty level than in the entire history of the country. Despite this statistic, we are still better off than most of the world.

We became the world leader through exploitation of our natural resources, our soils, water and forests. Today, these resources are at a critical stage and we must conserve them, recycle them and practice sustainable use. Most of all - we must change our lifestyles.

17. THE MOTHER EARTH SOCIETY

A Personal Note
Years ago my friend Bruce Boland and I were fervently involved with issues of the earth. We held seminars and meetings about the earth and its preservation. We had others interested in the cause and we referred to ourselves as The Mother Earth Society. We didn't form clubs or organizations or had membership lists. We simply touted preserving the earth. Other people spread the word and we knew that we had members, even if we didn't know them personally. Today, we still refer to ourselves as members of the Mother Earth Society and we hope you will consider yourself a member also. Here are the basic beliefs of our Mother Earth Society. We hope you will share them with us.

1. Mother Earth
Perhaps we can agree that a pot indicates a potter. So it is, we say that the existence of nature indicates the existence of God or at least what

we historically have thought of as God. Where do we go from there? No message from this God comes to us except we might speculate that we hear the message "you are here in this setting, make the best of it." It might be said that God is nature.

No one person has a direct link or communication with god anymore than any other. Perhaps a person who lives with and studies nature really does know more about God than one who does not study nature and natural processes. There is no prophet or messenger that knows God's Will any more than any other person. One who dwells on nature begins to understand the creation better than one who ignores natural systems.

Evolution of species is nature's way of changing and improving species. It is a fact supported by scientific studies of comparative anatomy, geographic distribution, DNA and paleontology. There are no contradictions in these studies. Those who disagree with evolution seem to have a "deeply" religious, rather than a scientific, background. That is, they arrive at their disagreement due to preconceived notions.

In teaching our youth about the world and nature, there is no need to dip into the realm of myths and legends and encourage fantasy. We can explain most of our existence on a scientific basis. Of course, there are things we can't explain, but there is no need to invent scenarios without labeling them as such.

Most adults have adopted the religion of their parents or the society in which they find themselves. To change that religion is a difficult undertaking. An educated person has studied history, science, literature, as well as other disciplines. If that person contemplates these studies and compares them with religious dogma, he can only be frustrated at the contradiction. The educated person realizes that no one religion has the monopoly on truth. Of course, in a society that is not free, it may be detrimental to one's health not to pretend to believe. In many Muslim societies women are forced to wear clothing that cover their bodies and men wear beards. Deviations from this practice leads to ostracism and punishment.

When youngsters have passed their teen years they have come to many conclusions regarding the value systems of the older generation and they, often as not, seek other value systems which results in rebellious behavior in dress, music and orthodoxy. I believe a lot of this is due to the contradictions presented by the lives of the older generation. In the United States we give socialism a bad name while more than half of our society operates under a socialistic system. We profess to adhere to the Ten Commandments and other supposed moral codes that have no relevancy in today's society.

There is no reason not to profess and teach a moral system outside the realm of religion. We generally know what is best for society and the individual, that is, we have some ideas of the needs for freedom of thought, movement, employment, democracy and independence.

2. A Dependence on Nature

All early humans eventually came up with the realization that they depended on Mother Earth for their sustenance. Once the realization of dependence on nature became finalized then it was a natural step to look for ways to enhance the bounty provided by nature. The bounty appeared to be controlled by unseen forces with jurisdiction over the sun, wind, rain, the seasons and so forth. The mind then set about trying to bargain with these supposed forces thus creating prayers, beseeching and sacrifices as well as talisman and amulets.

Today we know why the sun rises and sets, why we have the seasons, how it rains and the origin of winds. We know about solstices, equinoxes, eclipses, comets, planets and the like. We do not have to appease or bargain with unseen forces to perpetuate these.

The photos that have been taken in outer space depicts a mind-boggling array of images and substance. At this time we cannot conceive of its creation. In a way, it is frightening. Trying to fathom the intricacies of the substance of outer space gives us some idea of how the primitive mind tried to interpret the awesome world around it and came up with simplistic explanations.

What we do have is our home base, a fragile earth, and we are back to square one, the realization that we are dependent upon Mother Earth for our sustenance. We don't have to appease unseen deities. We can control our destiny by our actions.

We need to conserve natural resources, change our lifestyles, preserve the environment and limit population growth. Our religion should be to incorporate our scientific knowledge in the solution of our environmental problems and eliminate wishful thinking and the invoking of gods and goddesses.

We can use symbolism in rituals concerning gods and goddesses, **but we must not lose sight of the fact that gods and goddesses do not exist.** The use of ritual should be to emphasize our dependence on Mother Earth. The use of ritual can be to acknowledge the role our ancestors played in our creation and to foster a reverence for ancestors in future generations. Our worship should be of Mother Earth.

3

Most people have a feeling of brotherhood for their fellow humans and generally we wish each other well. However, when things go awry, evolution kicks in and competition and survival become our dominant force. We recognize this by enacting legislation to prevent the strong, or clever, from exploiting the weak. We do not wish to compete with each other on a scale that would annihilate one or the other, because it is vital to have harmony within our society and each member is important to society. Thus we have laws concerning slavery, murder, robbery, stealing and assault to which no conscientious citizen seriously objects. We also have laws concerning social behavior and activities which are often an area of debate.

We can agree on what is acceptable and what is not acceptable. Perhaps the objectionable behavior is simply an absence of education and understanding on the part of society, or perhaps a lack of education and understanding on the part of the individual engaged in that behavior. Generally, we wish individuals to conform to the norms of society. People usually conform. Violators are often people with economic or political power, and society usually catches up with them.

Good citizenship and social behavior are not the result of religious training. We all have references to very religious individuals who have violated the written and unwritten rules of society. i. e. priests buggering young parishioners, preachers involved in land scams, ministers involved in larceny. Religious leaders have the same attributes as other humans, since after all, they are human.

It is not religion which keeps society intact, but rather it is an innate knowledge within us of what is right and what is wrong. As members of society we must set rules that punish the serious offenders.

4

If we develop a philosophy of total preservation of earth environments we would certainly hinder development of our human culture. We cannot afford complete protection of pristine environments any more than we can afford their destruction. Some small native groups have been able to survive by living in harmony with the environment and only extracting basic needs, but with our burgeoning population we must exploit the forests, grasslands, minerals and soils. We can perhaps keep matters in balance with the practice of sustained yield for renewable resources and recycling for nonrenewable. In the case of soils, we work to prevent erosion and to replace depleted nutrients with fertilizers. We recognize nature and the natural environment as our religion and our temple but we cannot abandon their welfare to chance. **We are a part of nature, not apart from nature.**

5

We cannot deny that religion provides a model for social stability by reducing conflict, if all in the society accept it, and by promoting social control in the absence of legislation. Religion promotes a social ethic, at least an ethic for that society, and it brings people together in a spirit of communion and cooperation.

However, we cannot overlook the supernatural aspect of mainstream religion and their worship of deities. For this reason we might agree that **God is expressed in nature** which is the earth, and it is the earth that we should worship in our own way. This would give us all the benefits of religion and little of its distractions.

6. A Moral Guide

As members of society we must take responsibility for our actions. We must support individual freedom and become the enemies of those who would diminish our ability to make rational decisions concerning ourselves. We must reject the idea that supernatural influences have control of our lives. Beliefs and doctrines enforced by authority must be rejected, especially when they diminish our humanity. We can elevate our society by understanding that as human beings we are a part of nature and damage to the natural world is detrimental to the welfare of our position in the natural world.

It is an acceptable philosophy that the social function of present religions can be attained without a belief in deities or supernatural powers and influence. We can have rituals and dedication ceremonies without the invocation and benediction of the supernatural.

As much as we would like it to be true, there is no evidence for life after death and we must accept that. As our bodies decline and die there should be some comfort in knowing that our atoms and molecules will return to the universe to become some other part of nature. If anything, that is eternal existence.

Our goal should be to improve and preserve life on earth whenever we can. Those who choose to do otherwise should be eliminated from society.

7. An Earth-based Creed

There is no need for gods, demons, heavens or miracles in our spiritual life. We are on planet Earth, which is in a small corner of an apparently limitless universe. It is the earth that is important. Our religion and every effort we make in that direction should be to preserve the earth and keep it habitable for humans. We can use the entertainment value of symbolism and ritual to emphasize this commitment if we so desire.

8. Mother Earth

Those of you who agree that we humans are an integral part of this existence and believe we must work toward the protection of our earth are most welcome to take a place under the umbrella of Mother Earth. We know that Mother gave us life and continues to support us and we in turn must support Her, the Giver of all life. We recognize the sun as a continuing source of energy, which works with Mother Earth systems to support the fragile envelope of life surrounding our planet.

Sometime, in the future, you who take up this challenge may wish to add symbolism such as "the sun penetrates the earth and together they produce life, therefore the sun is masculine and the Earth is female." But let us not go overboard and assume gods and goddesses who demand worship. If you want gods and goddesses in your rituals, feel free to create them. Our basic worship should be appreciation of the earth and the preservation of its systems that promote life.

However, we do not discourage those who desire rituals in our veneration of the earth and its systems. Whatever brings us closer to an appreciation of the earth and its spiritual nature should not be discouraged.

We should frown on dogma or making anything but the appreciation of nature a prerequisite for a place under the loving umbrella of Mother Earth. When a census taker or pollster asks your religion give them Mother Earth. Perhaps in the future you will have The Sacred Church of Mother Earth, Friends of Mother Earth, The Mother Earth Society or some such name.

If we celebrate Mother Earth and work toward preserving and enhancing Her life giving support systems the world and its occupants will be many times blessed.

9. A Creed for Mother Earth We acknowledge
1. the role our ancestors played in our development
2. our place as part of the earth's ecosystem
3. we must protect our natural resources by conservation, wise and cautious consumption
4. other people have rights the same as we
5. we should treat people as we wish to be treated
6. we should be charitable to those less fortunate than we
7. sexuality is a part of being human
8. there are no gods or demons and we are responsible for our own destinies
9. we should refrain from activities which damage our habitat or that of others
10. we must strive to leave the world a better place in which to live.
11. an informed freely educated individual is in the best interest of society
12. that free inquiry is a basic human right
13. freedom of assembly to promote human welfare should be encouraged
14. censorship of ideas is repugnant

10. Environmental Ethics and Religion in Concert.

As an outdoor writer for more than twenty years, I often faced the problem of writing about hunter ethics. Most modem hunter ethics is determined by legal hunting regulations concerning species, sex, harvest numbers, days of harvest and areas of hunting. In societies where hunting is a necessary part of survival, the animals in question are almost always part of religious thinking. An animal will not be taken until the hunter has fulfilled certain ritual obligations, as well as moral obligations such as leaving breeding animals to survive and reproduce. In these religious societies there are certain rituals that must be followed in butchering and preparing the game. Failure to perform these rituals will most certainly doom future hunts.

Nomadic groups have kept their supply of animals constant by shifting hunting grounds from year to year, thus leaving an area build back to sustainable numbers. In many of these groups a spirit tells them when to move. Here perhaps, religion is evolved from logical reality.

We might look upon one religious belief among Hindu as questionable, and that is the idea of the "sacred cow." Upon examination it appears that this is really an environmental adaptation to survival conditions. In a land where starvation exists, taboo against eating cattle seems ridiculous at first glance, however, cattle roam freely and provide dung which is used as a fuel and fertilizer.

After continuous centuries of cultivation all arable lands need fertilizer. Thus the cattle supplies a needed commodity and survival for the greatest number. Cattle are also used to plow fields and to pull carts which are another aspect of the economy.

With earth systems everything is intertwined with everything else and we are all interdependent. We can never do anything to the earth that does not have consequences to other parts of the earth. If we spray for mosquitoes we deprive fish, bats and birds a source of food. So it is necessary to consider the consequences of our action. Once we have assessed the consequences and decide to continue the action then unless factors change we can engage in the action without further consideration.

Earth systems continue to change with variations in solar output, movement of the earth's crust and changes brought about by humans. We can adopt an attitude that we do not want change and try to keep the earth in its present state but change is a feature of the earth. Evidence of past changes have indicated a move toward improvement.

As changes occur, the biological communities tend to make modifications and adaptations. In some instances this might force species to move elsewhere as other species move into the new environment. Species can take a lot of environmental abuse but there are limits which must be observed.

The environment and each species within that environment can only tolerate a certain amount of change. Some changes will limit the number of a species in an environment. If you want more rabbits you have to do with less squirrels since they thrive in different environments. As trees take over grasslands the population of squirrels increases and the population of rabbits decreases. '

Species adapt to the temperature and precipitation of an area. Humans seem to be the only species that can survive in every environment. We are the only species to clothe ourselves and among the few species that can sustain ourselves by eating almost my organic matter.

Earth ecosystems survive because they take the energy from the sun and recycle it into different forms of energy. The Law of Conservation of Energy states that energy can neither be created or destroyed but only changed in form. Sunlight is changed into fossil fuels, trees and grass. We ourselves are storehouses of energy. The universe is one big bundle of energy.

There is no evidence of waste in nature. Organisms perish only to become nutrition for other organisms. Humans are part of this system. When we perish, the present practice of embalming prohibits immediate return back to the system.

Nature is a complex set of dynamics and it is difficult if not impossible to understand it all at our present level of development. The slow changes that are occurring are more than likely to escape our notice. This makes long range planning difficult.

One aspect of evolution is that over billions of years a variety of species have developed from the same roots. Species continue to develop and expand in numbers in response to the environment. The species that develop and are best fitted to the environment will survive. Those that cannot adapt will live out their lives but produce no viable offspring.

Comparative anatomy and measurement of body fluids identify the close relationships of one species to another. We are closely related to all other animals and plants on earth. No species can continue to grow indefinitely in size or in space. Earth has a limiting capacity and as humans, one goal should be to determine that capacity and stay within those limits.

Populations are limited by climate but other limiting factors may be the landscape and the nature of the chemical structure of the environment. One may be able to live in an area with a pleasant climate, but not if the area is subject to floods, volcanic action or other such events.

Our early pioneers looked upon nature as something to be dominated and conquered. They should have been looking for ways to become a part of that natural system. As we fit into the system we should retrieve our needs and wants but these should be attained with the least possible damage. We must get used to the idea that nature is not our enemy, but what sustains our very life.

When moves are made to alter nature we should take into consideration changes we might make to the integrity of biodiversity. We should think of future generations. How will this affect our great grandchildren, yet to be born.

We depend on other species for our own existence. Every species has a right to its existence or at least left alone to continue its existence without the overwhelming superiority of human inventiveness. However, we should be wary of species that would do us injury and take precautions in this regard.

Whether we continue to exist as a species is of no concern to the earth. However, the Earth is of great concern to us and we must treasure and protect it to the best of our abilities. The solar energy we receive is perpetual. We can expect it to continue for the foreseeable future but the same attitude is not true of the earth. If we don't recognize our stewardship of life forms, the earth might continue to exist, but not with humans on it.

There will be serious problems if we continue to increase in population beyond the capacity of our physical earth systems as they relate to earth resources. Lifestyles of the people living in Rotterdam are dependent on those who live along the Rhine River. They are at the end of the tube that is constantly being filled by others. The same may be said of those living near and around the Mississippi River Delta.

Waste production depends on resource consumption and we can alleviate both problems by limiting the creation of wastes. Limiting packaging and recycling will reduce the volume going into our landfills. Everything we are wearing or own at this moment will eventually end up in a landfill; so, it is necessary to reduce our consumption of unnecessary goods and make what we own last longer..

Conservation of energy and an energy efficiency philosophy will go a long way to protecting Mother Earth. We will always be dependent on energy for our progressing lifestyles, but think "why ride when 1 can walk."

Think of the earth as a bank and our use of resources as a withdrawal. We would not want to withdraw everything from the bank and have nothing left for the future. It is a healthy environment which guarantees our social, military, and economic security.

Earth depleting industries can be modified by substituting abundant materials for less abundant materials, substituting renewable resources for non-renewable resources.

When we act locally we should be thinking globally. As we fertilize our lawns the excess water runs into streams which run into rivers which eventually run into the oceans. Our contributions might be small, but the total amount accumulated in the stream is a multiplication of all of those who live in the vicinity of the stream. We are personally responsible for our own contribution to pollution or degradation of the environment.

Is our life more important than that of any other species? There are over six billion people on earth and less than two hundred whooping cranes. Excessive births of people lead to encroachment into new environments and excessive deaths of other species, as well as excessive deaths of people.

Mother Earth - Statement of Belief

We should worship Nature with the same zeal we have reserved for gods of the past. Since we exist because of Nature we recognize Nature as the source of all life. The sun sheds warmth on earth and Mother Earth responds by producing more life.

Our dedication and religion should be to the welfare of Mother Earth. All other religious belief is secondary. Those of us who need ritual in life can fulfill this need by observing the obvious sacred days of solstice and equinox as well as moon positions.

The spark of life which we call the soul is transferred from ancestors to us and we transfer this to our children. When we die, the spark of life ceases to exist and our substance returns to some other part of creation. Our spark of life continues in our children and their children'

There are two basic philosophies to the Mother Earth Movement, (1) that the Earth is sacred and we should all be concerned with Her welfare, and (2) all life is of divine origin. Therefore, we should venerate Mother Earth and, in the process, our ancestors who have passed their divinity on to us.

Walk lightly on the earth, listen to the songs of nature, the wind, the rushing waters, the pounding waves, the songs of birds and have holy-communion with Mother Earth. Strive to leave Mother Earth better than you found Her and Mother Earth will reward you and generations yet to come. - John Tomikel (written at Hawk's Nest)

www.ingramcontent.com/pod-product-compliance
Lightning Source LLC
Chambersburg PA
CBHW051542170526
45165CB00002B/850